MAKING THE MOST OF YOUR FARM BUILDINGS

A Guide for Farmers and Smallholders

MAKING THE MOST OF YOUR FARM BUILDINGS

A Guide for Farmers and Smallholders

R. W. LANGLEY

FOREWORD BY DR MIKE KELLY

THE CROWOOD PRESS

First published in 2006 by
The Crowood Press Ltd
Ramsbury, Marlborough
Wiltshire SN8 2HR

www.crowood.com

British Library Cataloguing-in-Publication Data
A catalogue record for this book is available from the British
Library.

ISBN 1 86126 846 7
EAN   978 1 86126 846 4

Disclaimer
Machinery, tools and equipment used on farms and in the
construction and maintenance of farm buildings should be used
in strict accordance with both the current health and safety
regulations and the manufacturer's instructions. The author and
the publisher do not accept any responsibility in any manner
whatsoever for any error or omission, or any loss, damage, injury,
adverse outcome, or liability of any kind incurred as a result of the
use of the information contained in this book, or reliance upon it.
Readers are advised to seek specific professional advice relating
to their particular farm building, or buildings, project and
circumstances before embarking upon any installation or
building work.

Typeset by Focus Publishing, Sevenoaks, Kent

Printed and bound in Great Britain by The Cromwell Press,
Trowbridge

Contents

Acknowledgements

I thank the following people and organizations for permission to reproduce photographs and figures: Mr Robert Ford (Brades Farm, Prees, Shropshire), Mr Stephen Craddock (Hill Farm, Agden, Cheshire), Mr Peter Langley (of Iscoyd, Shropshire), Mr Martin Ritchie (The Reid Rooms, Margaret Roding, Chelmsford), Mr K. Carlisle (Wyken Vineyards, Bury St Edmunds), Mrs C. Oakley (Rede Hall Farm, Bury St Edmunds), Blake House Craft Centre (Blake End, Braintree) and Mrs L. Shufflebottom (Shufflebottom Ltd, Llanelli). (The information in brackets that appears at the end of some of the picture captions is a shortened version of the bibliographical data concerning the publications in which the images previously appeared; the full details of these publications may be found in the list of references.)

I thank Dr Mike Kelly, a specialist with his own consultancy, Building Design, for his support and for contributing the Foreword.

I am grateful for permission to use a part of *Controlled Environments for Livestock*, written by the Farm Energy Centre at the NAC Stoneleigh, and for permission to use a part of *Farmwise – Your Essential Guide to Health and Safety in Agriculture*, written by the HSE.

Finally, I thank my family for their continued patience and encouragement along the way.

Foreword

Nothing contributes more to the content and convenience of a farmer than good and well disposed buildings. It elevates the mind, gives him spirit to pursue his operations with alacrity and contributes in many instances to augment his profits. On these accounts he ought to have them – I never yet saw a thriving tenant who had not neat houses.

This statement by James Anderson in 1794 is cited by David Soutar in his recollections of farm building design services in Scotland, written in 1999. Many of the greatest building and machinery innovators have been pioneering farmers, faced with a set of circumstances requiring radical thought, and to them agriculture owes a great debt; the loose housing of dairy cows in cow cubicles in the 1960s is a classic example of this.

The complexities of modern farm production systems demand more and more a team approach to farm building design. Animal health, welfare, nutrition and husbandry remain intimately linked to building design. The environmental impact of buildings, including their visual and pollution threats, is the subject of increasing legislation, as are health and safety issues. The role of the engineer is also becoming increasingly complex, to embrace new feeding, milking and handling systems, to name a few. This includes the sophisticated technology required to operate labour-saving innovations, such as automatic milking systems, feed dispensers and shedding gates.

But complexity of design is no defence if the basics are wrong. This book concentrates primarily on getting the basics right, ranging from constructional details to environmental control systems. The author has considerable experience as a lecturer specializing in agricultural buildings and equipment. Hence the text is well structured and will be helpful to those seeking information and an understanding of the basics of that information.

It is a pleasure to see a consolidation of some of the diverse information required to understand farm buildings brought together from an engineering viewpoint. This is helpful to both established farmers and new entrants into the industry as well as students and advisors. Farm buildings will continue to play a significant role in the development of British agriculture; but it is important that information sources keep pace with changes in the industry.

Dr Mike Kelly

A machinery shed constructed with steel, timber, concrete block walls and fibre-cement on the roof.

CHAPTER 1

Introduction

Over recent years the size and complexity of farm buildings has increased considerably; it is now fairly common to have hundreds of cattle in one building and perhaps 10,000 pigs or 100,000 poultry on one unit. Of course, this has brought with it problems of disease, infection and planning, to mention but three. The plain truth is that what would have been a viable unit a few years ago is no longer economic now. 'Thirty years ago a cereal farm of, say, 150 hectares or a dairy unit of 50 cows would each have been comfortably sufficient but neither could now be expected to provide a proper family income' (Prag, 2000).

The farmer's income, indeed, has fallen, and evidence points to further falls in the next few years and labour is also becoming less easy to hire. There may be more rewards by working outside agriculture altogether, certainly the financial one alone is well worth considering. There are many examples where a farmer's labour force is being drawn away from the land to operate under a much 'more acceptable' hours system and some important financial temptations too. (The author can vouch for one such system at least.) As one method to redress the balance, it may be worth considering putting up some well-designed farm buildings, even though the expense cannot be ignored, to offset these disadvantages; the labour saving can be hugely significant when the operating system is clearly thought out and planned precisely. The conversion of some existing buildings into feasible livestock units, for example, may also be justifiable in certain circumstances.

Of course, many farmers have taken the plunge and diversified some of their assets. Perhaps this is in the form of farm shops, residential conversion or business (office) uses. The success or otherwise of such enterprises will depend to a large extent on the proximity of roads (and the classification of them) to the farmyard. There are many farm buildings that are now redundant, perhaps owing to some farms having amalgamated, and something has to be done with them. In many instances diversification is the expected and almost hoped for lifeline; it may be the road back to some financial success (or simply to be able to operate 'in the black' for a change). Sometimes, when the old buildings are converted to alternative uses, this may pave the way for more modern and up-to-date structures to be erected for intensive livestock purposes. The old buildings may have low headroom, narrow doorways and be of a limited width, and will certainly represent more of a challenge when it comes to practical, modern farming. On the other hand, newer and large livestock sheds tend to be dimensioned according to current machinery uses and requirements.

Modern livestock buildings, especially those for cattle, may have some or all of the following features:

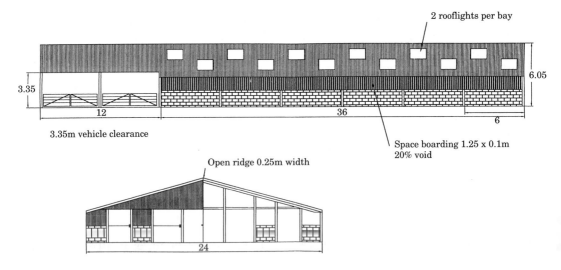

2 rooflights per bay

6.05

3.35

12

36

6

3.35m vehicle clearance

Space boarding 1.25 x 0.1m
20% void

Open ridge 0.25m width

24

*A layout of a modern farm building, showing the dimensions suited to
current farm machinery and typical livestock uses. It also allows for
maximum use of airflows for stock requirements.*

- wider, often without vertical posts as interruptions
- longer
- higher, usually with clearance at the eaves for a tractor with cab
- larger doorways to cope with tractors having safety cabs, together with larger towed equipment, for example, forage feed wagons
- often a steel structure fitted with a steel or fibre cement roof (sometimes with insulation), with a dwarf block/brick wall and Yorkshire (gapped) boarding on top of this
- usually fitted with concrete flooring to allow good hygiene and easy cleaning.

The justification for a new livestock building is really to provide a modification of the natural climate. If the climate around us were always 'perfect', then expensive buildings could be dispensed with, as they are, in fact, elsewhere in the world. However, in the United Kingdom the natural climate is only optimal for a short time; usually it is too hot, too cold, too damp, and so on. In the Tropics the necessary protection might be shading only, while in colder climates the need might be for more expensive and completely enclosed and controlled, ventilated buildings. In order to understand what we may achieve, the objectives of environmental control (in no particular order) within an agricultural building might be:

- to maximize (genetic) potential of livestock and birds
- to determine how much 'produce' will be produced under specific environmental conditions
- to determine whether the cost of modifying the environment will be compensated for by an increase in production
- to determine what types of environmental change are justified, and to what extent the natural climate should be modified.

We may wish to modify any or all of the environmental conditions inside a livestock building, such as the lighting, sound, thermal factors (such as temperature, humidity and air movement) and control of aerial contaminants. The thermal factors will involve heat transfer by means of conduction, convection, radiation and also vaporization (definitions will follow in Chapter 7). Livestock require quite narrow ranges of thermal environmental conditions for optimal performance, and stored produce requires even tighter conditions. Domestic animals are homeothermic, like humans, which means that they keep a constant body temperature at all times.

The effects of noise on livestock do not appear to have been investigated to any large extent. 'Where vocalisation is an important means of communication between animals we might expect some interference. For example, the noise of fans tends to disrupt the communication between sow and piglets at suckling' (Bruce, 1987). This is an environmental condition that will not be mentioned again, although perhaps it should.

On arable farms the use of older style buildings for grain storage is not normally considered feasible. The buildings are too low, too narrow and often have internal fixtures and fittings that are not conducive to modern techniques. Equipment has increased in size, and the demand for space in which to operate it has inevitably increased. There has also been a requirement for health and safety issues to be made more prominent, such as dust levels, lighting levels, ventilation and access/egress ladders. In days long ago these issues were not nearly so important in farm buildings, and, if one considers the conditions that used to exist on farms, it is a wonder that work used to be produced at all.

CHAPTER 2

Legal Aspects

There is often confusion surrounding the terms 'Building Regulations' and 'Planning Permission'. People may well realize that they require 'approval' before building new works, but they may be unsure as to which is which. In fact, they are quite separate pieces of building law, and quite often they will be dealt with by different people in different offices. It is likely that people will have to make separate applications for Building Regulation consent and Town Planning Permission. The application forms for these purposes are readily available from your local borough council offices; examples of these forms are shown in the Appendix (p.178).

THE LAW

During the Industrial Revolution of the nineteenth century there was huge shift of population from rural to urban areas and this created the first need for town planning. However, it was not until 1947 that the first Town and Country Planning Act was passed. There are many statutes governing planning law, but the main ones are the Town and Country Planning Act 1990 and the Planning (Listed Buildings and Conservation Areas) Act 1990. In addition, there are a number of pieces of secondary legislation (statutory instruments, known as Orders). Two important ones are the Town and Country

Planning (Use Classes) Order 1987 (as amended) and the Town and Country Planning (General Permitted Development) Order 1995.

Planning permission really deals with the correct use of land, the appearance of buildings, landscaping considerations, highway access and the impact that the development will have on the general environment. It is not the intention that protection will be given to one individual over another. The present position is that major works need planning permission from the council, but many minor works do not. Councils can use planning controls to protect the character and amenity of their area, while individuals have a reasonable degree of freedom to alter their property (Office of the Deputy Prime Minister, 2003). If planning permission is required it may be wise to wait until that permission is granted before proceeding to submit a building regulation application.

If you live in a house you can make certain types of minor change to it without needing to apply for planning permission. These rights, called 'permitted development rights', are derived from a general planning permission granted, not by the local authority, but by Parliament. In some areas of the country permitted development rights are more restricted. If you live in a Conservation Area, a National Park, an Area of Outstanding Natural Beauty (AONB) or the Norfolk or

Picture of old farm buildings no longer suitable for modern farming methods. They would, however, be suitable for some conversion.

The same buildings, they are of a traditional construction, of brick and slate roofing, and would be suitable for conversion, probably to some type of diversification.

Suffolk Broads, you will need to apply for planning permission for certain types of work which do not need an application in other areas. There are, in fact, three forms of development:

- permitted development which can be carried out without informing the Local Planning Authority (LPA)
- permitted development which requires the prior notification (officially termed 'determination') of the LPA before construction work begins (this is applicable only to agricultural developments)
- development which requires planning permission (NFU, 1994).

There may be other requirements if your house is a listed building (see later in this chapter).

The following useful summary is a checklist of which proposals will require planning permission:

1. *Full planning permission.* Is the proposed farm building:
   - on a holding of less than 0.4ha (1 acre)?
   - over 456sq m (545sq yd) in area including aprons and roads?
   - over 12m (39ft) in height?
   - over 3m (10ft) in height within 3km (1.9 miles) of an aerodrome?
   - within 25m (82ft) of a metalled trunk or classified road?
   - for intensive livestock, slurry or sewage, within 400m (438yd) of a protected building (a dwelling not linked to the farm)?
   - any works or buildings not for agricultural use?

If the answer to any of the above is yes, then full planning permission is required before the development

commences; contact should be made with the local authority.

2. *Prior determination for planning approval.* Is the proposed farm building:
   - less than 456sq m (545sq yd) in area?
   - a new or altered farm road?
   - a significant extension or alteration to existing with a cubic content of more than 10 per cent and any increase in height?

If the answer to any of the above is yes, then a Prior Notification application is required and should be made to the local authority.

3. *Building control approval.* Is the proposed farm building:
   - within 15m of (49ft), or 1.5 times the height, of a dwelling?
   - for retail, storage or exhibition?
   - for food processing?

If the answer to any of the above is yes, then a Building Control Approval application is required (in Scotland contact the local Building Control Office).

4. *Environmental Agency notification.* Is the proposed farm building:
   - for waste, slurry or effluent storage?
   - a silage clamp?
   - an agricultural fuel oil store?

If the answer to any of the above is yes, then the Environmental Agency should be consulted
(RMC, 2000).

In any planning application there is the potential problem of noise. This may be particularly acute in a quiet country setting even if the farmer has thought about any snags that may arise. Noise tends to travel further in the countryside compared with urban surroundings; for example, even if a clay-pigeon trap is sited in a remote place away from housing, there might be a noise disturbance sufficient to upset any planning decision. This may be particularly relevant in a 'diversification' plan.

It is as well to remember that a planning permission application is not only about the proposed building works but also whether the local traffic system can cope with the extra traffic. All schemes should be investigated thoroughly before planning applications are submitted and the work of building starts. Ideally, the planners should be invited to the site at an early stage, and their views ascertained. The planning process can often be a long drawn out affair; the pre-application stage typically takes around one to three months. In many instances it might be a good idea to invite the planning officer and county highways officer to the site itself in order for them to familiarize themselves with the plans. In some circumstances it will also be necessary to talk to the Environment Agency as well, if, for example, disposal from the site is going to be an issue. It is always advisable when an application is made that the widest possible use is specified, that is, B1 (offices), B2 (light industrial) and B8 (storage) (*Farmers Weekly*, 2 Feb. 2001).

Once the planning application has been submitted, it is a matter of waiting – a period of six to eight weeks is quite normal before the plans are put before a planning committee. The council must have sound reasoning in order that an application should be refused. The planning committee comprises ten to twelve people, all councillors from the district council, with many representing rural areas; one of them may well be the local councillor to the farm itself, so it might be favourable to invite them to a site visit as well. The committee will have a copy of the planning report, as compiled by the planning officer, and they will have this sent to every member of the committee a few days before the meeting. In fact, councils are legally obliged to let the farmer see the report three days before the planning meeting.

The planning committee meeting will run in the evening and often will go on quite late. The farmer who has put an application in is allowed to attend this meeting and permitted to put his case in a period of 3min. It is always advisable to attend this meeting, no matter how daunting it may sound to the average farmer since non-attendance is seen as a sign of disrespect. If the council refuses planning permission, the farmer then has six months in which to appeal to the Secretary of State; this might involve a written representation, a local hearing or a public enquiry.

The planning process may be given as an eight-point checklist (Prag, 2000), and is reproduced here as a summary aide-memoire:

1. Participate in local plan review
2. Check current local plan
3. Check whether planning permission is required
4. Hold informal discussions with, for example, local planning office and highway authority
5. Assess likelihood of planning permission being granted, in context of:
   • development plan
   • physical circumstances
   • suitability of proposal and location
   • access
   • sensitivity and potential objections

ABOVE: Converted piggeries now used as a banqueting hall and a wedding centre.

LEFT: The same farm, some of the old piggeries have now been converted to student accommodation.

6. Discuss with neighbours and (if appropriate) with councillors and pressure groups
7. Draw up plans
8. Submit application
   If consent granted:
   - check conditions and timescale
   - fulfil building regulations and other requirements
   - commence work
   If application refused:
   - assess grounds for refusal and consider whether possible to resubmit revised application
   - if not, discuss with advisers and consider possibility of appeal.

BUILDING REGULATIONS

These regulations (the Building Act 1984, with subsequent Regulations) deal mainly with health and safety matters in the interest of people who will use the building. In particular, they deal with:

- structural safety
- fire safety
- site preparation and resistance to moisture
- drainage
- toxic substances
- sound resistance
- ventilation

- hygiene
- drainage and waste disposal
- heat producing appliances
- protection from falling, collision and impact
- conservation of fuel and power
- glazing safety
- access and facilities for disabled people.

These are all legal requirements, but provided that the correct technical standards are met, the approval should be straightforward. It is the council's building control officers that deal with Building Regulation applications, and, as mentioned previously, this will be in a different office from the planning control officers.

Building Regulation approval will be required when you are:

- erecting a new building
- extending or altering an existing building
- changing the use of buildings
- installing or altering controlled services (including drains and some heating appliances and unvented hot water storage systems)
- installing cavity wall insulation
- re-covering a roof with different materials
- underpinning a building
- forming a structural opening (such as for a through room or new window opening)
- removing part of a chimney stack or chimney breasts.

The regulations apply to most buildings, including all residential, commercial, office, industrial and any farm buildings being converted for alternative uses. There are certain minor works, however, which do not need Building Regulation approval. These include some detached small garages, sheds, carports, porches, conservatories and other exempt buildings such as greenhouses and agricultural buildings. A building used for agriculture is one which is sited at a distance not less than one-and-a-half times its own height from any building containing sleeping accommodation, and is provided with a fire exit not more than 30m (98ft) from any point within the building. The definition of 'agriculture' includes horticulture, fruit growing, seed growing, dairy farming, fish farming and the breeding and keeping of livestock (including any creature kept for the production of food, wool, skins or fur, or for the purpose of its use in the farming of land). Agricultural buildings are not exempted if the main purpose for which they are used is retailing, packing or exhibiting (Powell-Smith & Bilkington, 1995).

It is always wise to approach the council, preferably in writing, to ask whether Building Regulation approval is required or not; in this way you will be able to keep a copy of the reply which may be important when you come to sell the property.

If your proposal does need approval, you may either make a Full Plans application, or submit a Building Notice. A Building Regulations Full Plans application is one in which plans, detailed specifications, and, if appropriate, structural calculations are given to the council. These would be checked to make sure that they meet the regulation requirements and, if they do, formal approval is then given. It may be necessary to give more information or to alter the plans; the council will give their decision within five weeks. A Building Notice is a written notice that you intend to carry out building work and does not necessarily mean that you must submit detailed drawings. More information may sometimes be requested to ensure that the proposed building is in line with the regulations. A Building Notice is more appropriate for minor residential alterations and/or extensions. No formal approval of plans is given and the work is assessed mainly through site inspection. You may start work 48hr after giving the Building Notice, although it is necessary to tell the building control officer when the work is being done to enable an inspection to be carried out. Both Building Regulation approvals and Building Notices are valid for three years from the date of application.

For a Building Notice you must supply:

• one copy of the form
• a block plan
• payment of the appropriate charge

and for Full Plans approval you must supply:

• two copies of the application form
• two plans and detailed specifications, and when appropriate, structural calculations or other engineering details
• extra copies of the plans called fire safety plans may be requested for non-residential works
• payment of the appropriate charge.

It may be possible for you to draw your own plans if you have the necessary training and knowledge; but, on the other hand, it might be more applicable to employ a professional person to act as your agent, especially if you intend to use the plans as a basis for obtaining builders' estimates, for instance.

By law, the building control officer must be told when the work reaches certain stages, the officer will then inspect the site. The stages are:

• commencement (48hr before commencement)
• excavation (24hr before concreting)
• foundation concrete (24hr before covering)
• oversite prior to concrete (24hr before covering)
• oversite concrete (24hr before covering)
• damp-proof courses (24hr before covering)
• drains (24hr before covering)
• drains, soil and vent pipes testing (not more than 5 days after completion)
• occupation before completion (not more than 5 days before occupation)
• completion (not more than 5 days after completion).

In practice, it is generally possible for an inspection to be carried out on the same day that notification is given, provided that it is received before 10 a.m. that day. Provided the works have a satisfactory final inspection by the building control officer, you will receive a completion certificate. This document will be useful to a property owner if he decides to sell it, and it may be required by solicitors and building societies, for example. It is worth noting that, if the correct notification is not given or the works do not comply with the Building Regulations, then the person responsible will be asked to open up or rectify the contravening work. As a last resort, matters may be referred to a magistrate's court for legal proceedings to begin. Action for alleged contravention of the regulations can be taken against the owner of the work and the builder; where proved, contravention of the Building Regulations is a criminal offence. (Summarized from 'Why do you need Building Regulations approval?', Chelmsford Borough Council, 2/99.)

BUILDING STANDARDS

Although there are no legal requirements to do so, most farm buildings will be built to the British Standard BS5502, essentially a code of practice. BS5502 (Buildings and Structures for Agriculture) was first published in 1978 and, since the original part on structures appeared, a further twenty-eight parts have been added to it. These are being continually updated and amended.

The Standard is divided up into five main sections covering general data, design, livestock buildings and ancillary buildings. Not only is BS5502 the country's 'yardstick', but it also has European and further international recognition; for example, the greenhouse section is presently being used as the base for producing a new EU standard.

When you receive a quotation for a new building, it is worth checking that it complies with all parts of BS5502. A completed building that conforms to all aspects of the structural requirements should have a plate on it stating that it does indeed comply. In addition, it should recommend certain limitations such as the maximum height of stored grain. All building drawings and written references should also make this information clear.

Classes for Agricultural Buildings (Part 22)

- Class 1. Unrestricted as to purpose and location, design life: 50 years
- Class 2. Not nearer than 10m (33ft) or zone of effect to a classified highway or human occupation not in the same ownership; human occupancy not normally greater than 6hr per day with the maximum density two persons per 50sq m (60sq yd); design life: 20 years
- Class 3. Not nearer than 20m (66ft) or zone of effect to a classified highway or human occupation not in the same ownership; human occupancy not normally greater than 2hr per day with the maximum density of one person per 50sq m; design life: 10 years
- Class 4. Not nearer than 30m (98ft) or zone of effect to a classified highway or human occupancy not in the same ownership; human occupancy not normally greater than 1hr per day with the maximum density one person per 50sq m; design life: 2 years.

It is worth noting that in the case of the design life this does not mean that, for example, a class 3 building is expected to fall down at the end of its design life. A small percentage of them would possibly not survive because the design loads that the building had been built to had been exceeded.

The flow chart shown overleaf may assist users in deciding which class of building will be most suitable in their situation.

Some typical questions might be: what class of building is required for 3hr human occupancy, one person per 50sq m (60sq yd) floor area, 11m (36ft) from a highway or dwelling? Answer: class 2, 20 years life. And what class of building is required for 1.5hr human occupancy, one person per 50sq m floor area, 8m (26ft) from a highway or dwelling? Answer: class 1, 50 years life.

OTHER LEGAL ISSUES

Covenant

This is a contractual agreement whereby one party, the covenantor, agrees to do or

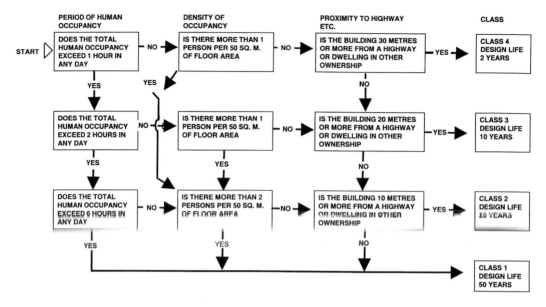

A flowchart used to select the class of building (to BS5502).
(Barnes and Mander, 1991)

not to do something for the benefit of the other, the covenantee. The covenantor has the burden of the covenant, while the covenantee has the benefit of it. A covenant will normally be made in a conveyance of freehold land or in a lease. Examples of covenants which might be entered into between the parties to a freehold conveyance include covenants preventing building on the land conveyed, preventing the building of more than one house, limiting the property to residential use or requiring the building of a boundary wall. Getting someone else's agreement may be necessary even if you do not need to apply for planning permission, and this may be checked by consulting a solicitor.

Listed Building Consent

This will be required when you want to demolish a listed building or you want to alter (internally or externally) or extend a listed building in a manner which would affect its character as a building of special architectural or historic interest.

Listed building control is in addition to that over the development of land. You may also need listed building consent for any works to buildings within the grounds of a listed building. It could be that the painting of a front door is subject to listed building control; conversely, simple cleaning work to a listed building would not normally require consent, but it is wise to check the position carefully with the council before undertaking any works. Note that it is a criminal offence to carry out work which needs listed building consent without obtaining it beforehand.

There are a number of factors which are relevant to the consideration of listed building applications to the local authority. These are:

- the importance of the building, its intrinsic architectural and historic interest and rarity
- the particular physical features of the building which justify its inclusion in the list
- the building's setting and its contribution to the local scene, which may be important
- the extent to which the proposed works would bring substantial benefits for the community, in particular by contributing to the economic regeneration of the area or the enhancement of its environment (including listed buildings). (DETR, 1994)

Any building that is listed will be given a certain grading and this may be one of the following: Grades I and II* identify the outstanding architectural or historic interest of a small proportion (about 6 per cent) of all listed buildings; Grade II includes about 94 per cent of all listed buildings, representing a major element in the historic quality of our towns, villages and countryside. (DETR, 1994) A farm building might therefore have a listing on it, largely depending on its age; this is certainly the case if the building is older than 1700. Most buildings put up between 1700 and 1840 are also listed. If the building dates from between 1840 and 1914 only those of 'quality' will be listed, and, after 1914, only buildings judged to be of a particularly high quality will be listed.

The main responsibilities involved in owning a listed building may be summarized as:

- repairs notice; the LPA will serve one of these if they consider the work is necessary for the proper preservation of the building
- urgent work to preserve listed buildings; the LPA may carry out work themselves if they consider it is required to keep the building in a good state of repair, but this will be done only if the building is unoccupied
- compulsory purchase; the LPA can use their powers to buy a property which they consider is in need of repair.

Conservation Area Consent

This may be required if you live in a conservation area. It is an area of special architectural or historic interest, the character or appearance of which it is desirable to preserve or enhance. Anyone proposing to cut down, top or lop a tree in a conservation area is required to give the local planning authority six weeks' notice. In particular, you will need consent to do the following:

- demolish a building with a volume of more than 115cu.m (150cu.yd); there are a few exceptions to this general rule
- to demolish a gate, fence, wall or railing over 1m (3ft) high where it is next to a highway (including a public footpath or bridleway) or public open space; or over 2m (6ft) high elsewhere.

An example application form for work to be done on a listed building/conservation area is shown at the end of the book.

Environmental Impact Assessment (EIA)

Planning authorities will decide on the necessity for an Environmental Assessment and will inform the applicant soon after submission of the full planning applications. They must inform the applicant within three weeks of the planning application being submitted. The objective is to prevent the creation of pollution or nuisance at source for large-scale proposals requiring planning consent. For intensive livestock systems, the following thresholds apply: broilers – more than 85,000 places, hens – more than 60,000 places and sows – over 800 places. In all these cases an EIA is mandatory. Below the above levels an EIA will be required at the discretion of the local authority, although the perceived environmental effects must be significant and a guide of 500sq m (600sq yd) floor space is used. There are also requirements for irrigation, water storage and unimproved grassland.

Tree Preservation Orders (TPOs)

These may well be present. This means that many trees are, in fact, protected by a TPO. In general, this means that you will need the council's permission to prune or fell them; again, it is a criminal offence to undertake this work without prior consent.

Rights of Way

These may be yet another restriction on your proposed development. If the building is going to obstruct a public footpath which crosses your property, you must discuss the plans with the council at an early stage. The granting of planning

Converted piggeries now used as a banqueting hall and a wedding centre. Note that sensitive planting around buildings can add a sense of permanency and belonging and will certainly assist in the planning process.

permission will not give you the right to interfere with, obstruct or move the path. A path cannot legally be diverted or closed unless the council has made an order to do so to allow the development to go ahead. The order will be advertised and anyone may object. The owner may not obstruct the path until any objections have been considered and the order has been confirmed. It is worth bearing in mind that confirmation is not automatic; for example, an alternative line for a path may be proposed. Planning permission for a new gate would not itself grant you any right of way on land outside your own.

Advertising

This is another matter where you will need to apply to the council before displaying an advertisement bigger than 0.3sq m (3sq.ft) on the front of, or outside, your property. A sign no larger than 0.3sq m, if it is for identification, direction or warning, such as your house name or number or 'Beware of the dog', would not require this consent. Temporary notices of up to 0.6sq m (6.5sq.ft) relating to local events, such as fetes and concerts, may be displayed for a short period without your having to apply to the council. There are different rules for estate agents' boards, but, in general terms, these should not be bigger than 0.5sq m (5sq.ft).

Wildlife

This should also be considered when planning to execute major works. Animals, plants and habitats may be protected under their own legislation (for example, badgers), under the Wildlife and Countryside Act (for example bats, which sometimes roost in houses) or under European legislation (EU-protected species such as the great crested newt). English Nature may be able to provide advice on what species are protected and what course of action is advised. Note that, even when your development proposal benefits from permitted development rights, the legal protections for wildlife still apply. (Based on extracts from the publication: Office of the Deputy Prime Minister, 2003.)

ABOVE: A converted set of farm buildings now in use as a farm shop.

LEFT: The same farm, the buildings used for 'business' are more or less standard farm buildings, but extensive repair work has been carried out.

BELOW: The same farm, all the buildings have been re-roofed.

CHAPTER 3

Traditional and New Farm Buildings

Traditional farm buildings tend to blend harmoniously into their landscapes. This is due to the materials that were used in their construction and the traditional methods used in both farming and building work. Together these tended to limit the scale and the form of the buildings to something approaching domestic proportions. Additionally, the colour and texture of materials were restricted to pleasant shades, and the siting and positioning done to gain shelter and respect topography. Depending on the part of country under consideration, the choice of materials and techniques will vary considerably from the selection of available timber, stone and brick. The orientation of the building is an important factor to consider. Open-fronted buildings usually face south or south-east, whereas enclosed stock buildings are often best on a north-south axis so that both flanks receive sunlight, either in the morning or the evening. The prevailing wind direction may also have a bearing here.

Modern buildings and traditional ones are completely different, particularly due to their size, scale and industrially-produced materials. The newer buildings may represent quite a 'planning' challenge in hilly country, where they may be visible over wide areas; ways must be sought to help them to fit into the landscape without compromising their impor-

tant functions. Location and siting become important in the initial design considerations. As a general rule, new buildings should be sited away from prominent hilltop or skyline locations, and should preferably be positioned in hollows where there is much more chance that topography and trees will provide natural cover. In areas where hills are common, it may be preferable to build on sloping sites to give respect to the natural contours. When building near to existing structures, the general advice is to make the new 'respect' and be 'sympathetic' in terms of design, scale and finish to the old; new structures should not stand in bold isolation to them. The loss of hedges and trees has increased the number of views across the landscape, although perhaps this problem is not as acute now as it was several years ago.

Many modern buildings are of single-span, low-pitched portal frame design and do not blend particularly well with older forms; they are large and tend to break tradition and they are either much bulkier or tend to be long, low sheds. The services are also very prominent, such as electrical supply, extract fans and ducts, conveyors and feed hoppers. However, the planning object is to reduce the apparent scale and bulk of the building and to improve on its form by, for example, reducing wide, single-span buildings into

Roof overhang reduces scale.

Scale will appear greater if the same
materials are used for both roof and walls.

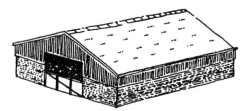

Roofs should always be darker
than walls.

Sketches of farm buildings in the landscape. Certain features will make it
appear better or, in many cases, worse. (Cull, 1987)

multiple-span units of smaller size, the
alteration of roof over-sail, the choice and
colour of materials and the use of additive
rather than subtractive forms – all can
help to make buildings fit in better. As one
rule to bear in mind, the 'all-under-one-
roof', rectangular structures idea is one to
avoid as much as possible. The roof pitch
should reflect existing buildings and
topography, that is, generally roofs tend
to be steeper in hill areas and shallower
in flat countryside. See the accompanying
figures for further guidance.

The colour of building components is
most influential in building design.
Large, modern buildings of natural
'asbestos' sheeting, which is very bright
when new and drab when weathered, are
visible for great distances in the countryside. As general colour guidance, the
British landscape is not one of harsh
colours; traditionally roof coverings are
darker than walls and dark colours visu-
ally recede and light colours assert themselves. Of course, as the predominant
geology changes around the country, so
too does the advice on building colours.
For example, in areas where natural

limestone is prevalent, white roof sheets may be perfectly acceptable; in other areas dark, warm colours, such as brown, dark green and dark grey, may be more applicable. Local design guides produced by councils will be able to advise on the most suitable choice of colours for specific areas. Careful choice can reduce the apparent scale of a large farm building; darker roof coverings reduce scale and ensure that a building with a darker roof will fit better into its surroundings than one clad in bright, coloured sheeting. It is generally thought best to clad over the stanchions in order to further reduce scale.

A sketch showing how natural features can reduce the sense of scale of the building. (Cull, 1987)

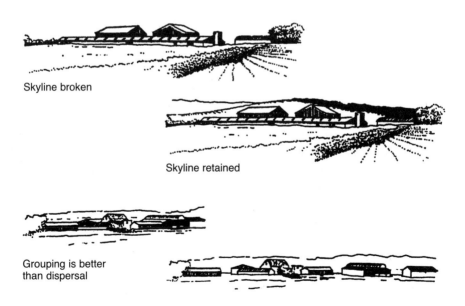

Skyline broken

Skyline retained

Grouping is better than dispersal

Sketches showing how to 'fix' the surrounding landscape and how to group farm buildings correctly in order to reduce the scale. (Cull, 1987)

There are several ways colour can be used to good effect:

- plastic-coated steel, available in a wide range of colours
- factory-applied paint finishes to cement-fibre cladding
- stained, creosoted or plain timber boarding allowed to weather naturally
- brickwork, available in a wide range of textures and colours
- bitumen or other colour finishes to concrete blockwork.

Advice on the materials chosen really concerns texture and colour. Texture is the shadow cast by the interruption of the surface, and this is important because many modern finishes have highly reflective surfaces and these, of course, can be seen from quite a distance. This means that the careful use of space boarding, recessed, pointed concrete blockwork, brickwork and corrugation pitch in sympathy with building size are important factors to consider. The texture and use of materials is vitally influential when a new building is to be located near to some other perhaps traditional buildings.

As regards detailing, perhaps the most important aspect is roof over-sail. A relatively simple eaves and verge detail, that is, making the roof over-sail, creates a shadow and has a marked effect in defining the roof and wall elements. The overall effect is to reduce the bulk of

Utilise existing walls wherever possible.

A stepped construction on a sloping site helps break up an otherwise unbroken roof line.

These diagrams show how the use of surrounding features, such as walls, and also a stepped construction technique may assist in reducing the impact of the buildings. (North Pennines Area of Outstanding Natural Beauty Steering Group, 1998)

Isolated buildings should, where possible, take advantage of natural dips in the land.

RIGHT: Two diagrams showing how a single building may successfully be situated, and in another case where the choice of location is unfavourable. (North Pennines Area of Outstanding Natural Beauty Steering Group, 1998)

In prominent positions, modern buildings can have a significant impact on the landscape.

A group of buildings all have parallel roof lines, which helps to reduce the visual effect.

ABOVE AND RIGHT: These sketches show how parallel roof lines help to reduce visual impact, and how the choice of the appropriate site can also assist. (North Pennines Area of Outstanding Natural Beauty Steering Group, 1998)

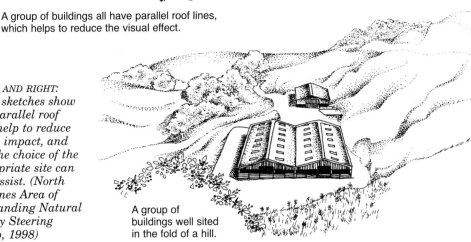

A group of buildings well sited in the fold of a hill.

the building and in so doing create visual interest. The eaves can be extended by 0.4–1.2m (1.3–4ft) to create shadow lines. This is one feature that has been adopted successfully on some domestic building designs and is an aspect that possibly ought to feature more in agricultural building design. It is generally favourable to avoid near flat, monopitch roofs, their appearance is not good and they are unlikely to withstand high wind and snow loads or be completely waterproof, and ventilation does not work well under them.

Landscaping, in particular by planting trees, around a new building can help to soften a hard outline, perhaps break up a silhouette or assist in providing an 'anchor' for the building in question. Fast-growing trees such as evergreen cupressus or Lombardy poplar are not native, usually do not blend in with the landscape and can often draw attention to the very building they are trying to hide. Sometimes a mixture of broadleaf and coniferous species adds variety, with the latter acting as a nurse crop for the slower growing, native hardwoods. It is always best to use native trees similar to those in the neighbourhood. Shrubs and climbing plants that may look attractive on a house wall will look equally attractive on a farm building and establish more quickly than trees. It should be remembered to plant trees as far away from a building as their mature height (as a rule of thumb) in order to avoid any future root damage to their foundations. Never plant trees close to drainage pipes.

It is wise to retain as much of the existing landscape features as possible when planning new farm buildings. It is often said that the use of natural and newly planted features can have the effect of reducing the apparent scale of the buildings. The use of cut-and-fill techniques and earth banking is another recommended way to make new buildings fit in with a hilly landscape. Once the building has been constructed it is necessary to regrade the surrounding ground to blend in with the existing landform; to restore the top soil and reseed to give a well-cared for appearance.

CHAPTER 4

Constructional Details

CONCRETE

The earliest known concrete dates from around 5600BC and formed the floor in a hut in the former Yugoslavia. The cement used was red lime, probably from 200 miles upstream, and this was mixed with sand, gravel and water before compaction. It has recently been suggested that the pyramids themselves were made from cast concrete blocks and not cut stone. The name concrete comes from *concretus*, a Latin word meaning bringing together to form a composite. Even admixtures such as blood or milk were used in the past to give the concrete certain qualities. After the fall of the Roman Empire the skills to make and lay concrete were lost until around 200 years ago.

Present-day Use

The main advantages of cement-based materials are seen as: low cost, flexibility of application (for example, mortars, concretes and grouts), variety of finishes obtainable, good compressive strength and protection of embedded steel. And the following represent some disadvantages/problem areas: low tensile strength/brittleness and rather high density (although lower density types are available), susceptibility to frost/chemical deterioration (depending on type) (Taylor, 1997).

The mixes that can be used for different applications are shown overleaf.

Admixtures

Admixtures are material added during the process of mixing concrete in a quantity of less than 5 per cent (by mass) to modify the properties of the mix in the fresh and/or hardened state. They are not substitutes for good practice though and are unlikely to make poor concrete better. With their cost being minimal, they are usually seen as something that is worthwhile. The main types available are:

Water Reducing/Plasticizing
This causes the individual cement particles to become better dispersed in fresh concrete so that its fluidity is increased. The result is that it gives enhanced workability, hence better compaction without more water. Or it can give the same workability with less water, hence giving stronger, less permeable, hardened concrete. It can be useful with rich mixes used in site construction as these can be difficult to place. They are usually limited to 0.5 per cent by mass, and can give a water reduction greater than 5 per cent at constant workability, or increase in slump of up to 100mm. They can be used with slag, PFA and OPC (PFA is pulverized fuel ash; OPC is ordinary Portland cement).

Application	Finish	Agricultural Requirements	Quality Assurance & Regulations	Benefits	Notes
Readymix Stockfloor					
Parlour	• wood float • brush	• cleanable • non-porous • crack and crevice free • reduced acid attack	• milk and dairy regulations • milk buyers QA contract	• hygienic • cleanable • low permeability • fibre benefits	fibre + stockfloor = dense, strong, low porosity and durability
Cubicles	• wood float	• cleanable • non-slip • abrasion resistance	Welfare regs: free from stress and injury	• hygienic • cleanable • low permeability	
Bedded Yards	• heavy brush	• cleanable • non-slip • abrasion resistance	Welfare regs: free from stress and injury	• hygienic • cleanable • low permeability	
Livestock Passages Traffic	• internal-brush • tamp or groove • external-tamp	• non-slip • abrasion resistance levels • washable	Welfare regs: free from stress and injury	• fibre benefits • brush= less latiance • less new concrete disease • durability • abrasion resistance	tamped finishes which are regularly scraped will become smooth in time
Vehicle Traffic Internal	• tamped	• grip/non-slip		• fibre benefits • crack resistance	
Readymix Multistore					
Grain Straights (Animal Feed) General Storage	• wood float • power float • not polished	• crack free • abrasion resistance • cleanable • low porosity	Assured Combinable Crops Scheme (ACCS)	• low permeability • easily cleaned with brush and vacuum • vehicle grip	polished finishes are too smooth
Readymix Liquitite					
Slurry Storage	• wood float • power float	• impermeability	Control of Pollution for Agriculture regs (COPA)	• low permeability	
Silage Clamps	• tamp or brush	• impermeability	Control of Pollution for Agriculture regs (COPA)	• low permeability	

A table from Readymix showing the applications and advantages of different mixes for each typical use. A stockfloor, a multistore and 'liquitite' are considered. (Cemex Readymix, 2005)

High-range Water Reducing /
Superplasticizing
This has the same effect as the above, but more so. They can be used by up to 3 per cent by mass of cement and water reductions of over 12.5 per cent (up to 30 per cent) can be achieved and a 'collapsed' slump reached. They are used for slag, PFA or OPC. The fluidity of flowing concrete is such that virtually no compaction is required, thus floors should be perfectly level and formwork watertight. The effect of these will normally last for about 30min. On the downside, these are quite expensive and any possible benefits must be calculated carefully.

Water Retaining
These admixtures reduce the loss of water by bleeding. They are rarely used in Britain for concrete but often added to mortars. They reduce the suction effect of masonry units, which leads to rapid stiffening of the mortar. The admixtures are often specified for rendering mortar and for autoclaved, aerated concrete blocks. They may be used for a pumping aid and underwater concrete.

Air Entraining
This allows a controlled quantity of small, uniformly distributed air bubbles to be incorporated during mixing, which remain after hardening. These give space for ice to expand into when water freezes, thus preventing damage to concrete. It is particularly good when attack is aggravated by de-icing salts used in winter. They also give a good cast surface, together with reduced plastic shrinkage, settlement and bleeding. In Britain we use about 4 per cent air, whereas in the USA 8 per cent is more common. With mortars, around 20 per cent air is used. Entrained air reduces the compressive and flexural strength of concrete, for

example, a loss in strength of 5.5 per cent for each 1 per cent of entrained air. But air bubbles increase workability, allowing a reduction in water content which will offset the strength loss otherwise occurring. It is useful to know that the mixing up of ordinary concrete is unlikely to entrain more than 1 per cent air (ignoring air voids caused by poor compaction). Air entraining should be used for all external pavings on the farm.

Set Accelerating
This decreases the time to the start of the transition of the mix from a plastic to a rigid state. These agents may not affect the rate of strength development. They are not common in Britain, except for when used in sprayed concrete. As an example, the setting time at 20°C could be reduced by more than 30min. The compressive strength should be more than 80 per cent of a control mix at 28 days. They can be used as liquids with the water or as solids.

Hardening Accelerating
This admixture increases the rate of development of early strength of concrete, with or without affecting the setting time. Even at low temperatures, they give early strengths comparable to concrete at 20°C, for instance, a 24hr-strength should be more than 120 per cent of that of the control mix at 20°C. Hardening accelerators cause a reduction in later strength, but must be more than 90 per cent of the control at 20°C and 28 days. They are often referred to as 'anti-freeze' and in high doses allow bricklaying to proceed in freezing conditions; this is not acceptable practice. Concrete must still be protected from frost in the normal way.

Set Retarding
These agents extend the time of the start

of the transition of the mix from the plastic to the rigid state. In other words, they prolong the workable life of concrete or mortar. In warm weather when small gangs are handling large loads in awkward conditions, this attribute may be useful. Early strengths are reduced, but later strengths are enhanced. Most retarders have some plasticizing properties in addition. Typical retarding times used might be: 1–2hr for concrete, 8hr for floor screeds and up to 72hr for ready-to-use mortars. Heavy metals can cause retardation; impurities in aggregates, for example, can cause problems.

Water Resisting
These materials reduce the capillary absorption of hardened concrete. Sometimes they are referred to as 'water-proofers' but this is not a good name and can be misleading. The effect seems to be to coat the inner surfaces of capillary pores with a water repellent, so reducing the passage of water through the concrete. They may also reduce lime-bloom and efflorescence of concrete bricks and blocks. Water-resisting admixtures are used in pre-cast concrete but rarely in mass concrete; they are widely specified for rendering mortars.

Multifunctional
These affect several properties of fresh and/or hardened concrete by performing more than one of the functions above. Typical examples in Britain are retarding and accelerating water reducers and air-entraining plasticizers.

Other Admixtures

There are, of course, other types of admixture and these may include: admixtures for foamed concrete (flowable fill), corrosion inhibitors and underwater concrete. Yet other materials may be supplied by the admixture industry, often incorporated in excess of 5 per cent by mass. These may include silica fume, fibres, pigments, polymers, latexes, mould-release agents and wash-water systems.

Concrete Ingredients

The basic concrete ingredients are cement, sand (or crushed rock fines, not now referred to as fine aggregate), stones (either gravel or crushed rock, also called coarse aggregate) and water. There may also be reinforcement which might comprise rods of steel or fibres of steel/polypropylene/glass. These may not be used for strengthening but primarily to prevent subsequent cracking in the concrete. The quantities of the ingredients must be carefully controlled to produce a consistent quality of concrete. By altering the proportions of the several ingredients the primary qualities of concrete can be changed, that is its strength, abrasion resistance and chemical resistance.

Properties of Fresh Concrete

These may be briefly stated as:

* workability, that property of a plastic concrete mixture which determines the ease with which it can be placed and the degree to which it resists segregation; it embodies the combined effect of mobility and cohesiveness
* stability, the ability of the mix to remain homogeneous, resisting segregation and bleeding
* compactability, the ease and amount of

void reduction that can be achieved when concrete is compacted

- spreadability, the ease by which concrete spreads when subjected to vibration.

Properties of Hardened Concrete

There are really only two properties to consider under this heading:

- strength, this is normally considered to be the most important property in relation to mature concrete and usually means compressive strength, as measured by the cube test; strength may be affected by:
  - free water/cement ratio, the more free water there is in relation to cement, the lower the strength
  - aggregate properties, crushed aggregates result in higher strength than uncrushed, since they form a better key with hydrated cement; to obtain very high strengths the use of crushed aggregate may be essential
  - the cement type
- durability, this depends on the permeability and hence the porosity of hydrated cement, that is, a low water/cement ratio equals greatest durability: 0.4 is a good durability, 0.8 is a poor durability.

CEMENTS

A Joseph Aspdin invented cement, as we know it today, in 1824. He produced a powder made from a calcined mixture of limestone and clay and called it Portland cement because, when it hardened, it produced a material resembling Portland stone from quarries near Portland in Dorset, but importantly, it was cheaper than the stone. Portland cement is produced by igniting a mixture of two materials: one rich in lime – such as limestone or chalk and one rich in silica – such as clay or shale. The main constituents of cement can be given as (in percentages): calcium oxide (60–70), silica (17–25), alumina (3–8), iron oxide (0.5–6), magnesium oxide (0.1–4), alkalis (0.2–1.3) and sulphur trioxide (1–3). It is well to remember that Portland cement does not describe a unique product; there are many variations. The relevant standards, which will not be covered here, lay down the limits of its properties. The standard bag size is either 50kg or, increasingly commonly, 25kg in most DIY stores.

Types of Cement

1. ordinary Portland cement (OPC); this is used for the bulk of farm concrete concrete/mortar; it is not now called OPC but rather PC
2. rapid-hardening Portland cement (RHPC); this is used when high early strength is required, it is finer in its bulk state
3. masonry cement; this is used in mortars but not in concrete itself; it has admixtures included at the factory
4. sulphate-resisting Portland cement (SRPC); this is made by adding iron compounds during the kiln process to produce a material less affected by acid waters and other injurious salts, that is, it is better in areas where sulphates are found naturally in the ground
5. low-heat Portland cement; this is used for massive concrete pours, such as dams, to reduce the heat of hydration generated during the chemical

reaction; it will be appreciated that to accompany any chemical reaction there will usually be heat given off.

Cement Replacements or Additions

For technical and/or economic reasons, it may be appropriate to replace part of the cement content of the concrete mix by, for example:

1. ground granulated blast furnace slag (GGBS), a by-product of the steel industry and available in only some parts of the country; when used with Portland cement it forms Portland-blast-furnace cements; a British Standard exists for this product and it has been used in civil engineering projects for many years, producing an improved sulphate resistance; when in use up to 65 per cent replacement of cement is permissible
2. pulverized-fuel ash (PFA or fly ash), a by-product formed by burning pulverized coal at power stations; fly ash is precipitated from the flue gases; quality varies with the type of coal used and the conditions of burning; not all PFA is suitable in concrete – some has qualities that are not useful; the material may be interground or blended with PC to form Portland pulverized-fuel ash cements, 15–35 per cent PFA may be added this way; alternatively, PFA can be combined with cement by ready-mix concrete companies, the customer will then need to make the decisions
3. silica fume/micro silica, a by-product of the electric arc furnaces that used to manufacture silicon and ferrosilicon; most material is imported from Scandinavia and Canada and it is

strongly pozzolanic and shows good potential for cement replacement; it shows improved resistance to sulphate attack.

There are other cement replacements that may be used, for example, calcined clay and metakaolin – a product of china clay. These are specialized materials and not particularly suitable for agricultural applications.

Effect on Concrete of Using Cement Replacements

The following properties may be affected:

1. rate of gain of strength: concrete with slag or fly ash gains strength much more slowly than concrete without them, that is, need to wait longer between laying and using the concrete
2. curing: this is explained in more detail below; it is even more important to do this properly when working with cement additions
3. frost resistance: this is similar to ordinary cement, however, it is important to note that frost resistance is greater when strength is higher and the rate of gain of strength is lower if slag and fly ash are used
4. air-entrainment: this is recommended for all farm roads and similar projects; slag can help to produce concrete that is resistant to a combination of frost and salt, with fly ash high doses of admixture may have to be used and there might be problems in controlling the amount of air in the mix
5. working: concretes with slag and fly ash may have higher workabilities or higher slumps and may be easier to place and compact; they have increased

setting times, which may work out to be an advantage in hot weather but a disadvantage in cold weather

6. durability: concretes have greater resistance to sulphate attack, but are not as good for this as sulphate-resisting Portland cement.

Cement/Water Reaction

Portland cement is said to be hydraulic, that is, it sets hard by the action of water only. In other words, when water is added to cement a chemical reaction occurs, called hydration. Heat is given out and the resultant new chemical binds coarse aggregate and sand together; only a small quantity of water is required for this process. In fact, the water in the atmosphere is enough to cause 'air set'. Thus it is essential to keep cement dry – a polythene inner bag is now used to keep moisture out. It is important that cement bags are kept off the floor and stored in a dry building; in many cases it may be wise to wrap the cement bags in another polythene bag for added protection. An unopened bag may be stored for up to 6 weeks.

In theory, only 25kg of water are needed per 100kg of cement to complete the reaction. However, in practice at least 40kg of water are required and sometimes 100kg of water to 100kg of cement. Extra water is required to make the concrete workable so that it can be placed and properly compacted. Note that the aggregate contains an unknown amount of water and sometimes, when the aggregate is stored outdoors perhaps for a long time, this can have quite an effect on the mix. But the more water that is added (above that required for good compaction), the weaker and less durable the concrete becomes.

Water

The water used in concrete mixes must be clean, that is, potable. Dirty, contaminated water will reduce the strength and durability of the concrete and salty water, such as sea water, should be used only in unreinforced concrete and in an emergency. It is imperative that salty water must never be used in reinforced concrete, if it is you can get severe corrosion of the steel.

AGGREGATES

Aggregates are used in concrete for the following reasons: they greatly reduce costs, reduce heat output and hence reduce thermal stress, reduce shrinkage of concrete and help produce concrete with satisfactory plastic properties; the type used will also affect the strength of the concrete (from Taylor, 1994).

There are many other types of aggregate that may be used for special reasons such as:

- low-density concrete
- high-density concrete (primarily for radiation shading)
- abrasion-resistance concrete, such as floors (granite or carborundum aggregate)
- improved fire resistance
- decorative aggregates.

But whatever aggregates are used, they should be from a recognized supplier and therefore be clean and free from constituents that decompose or change significantly in volume such as organic matter. They must also be free from constituents that react with cement such as sulphates. If possible, it is always wise to use aggregates that comply with recognized standards. Normal aggregates are 'dense' ones, such as sand, gravel or crushed rock. It is possible to obtain man-made, lightweight aggregates and these are commonly used to make concrete blocks and for insulated floors, for instance, expanded perlite and exfoliated vermiculite.

The grading of aggregates, that is the relative proportions of different sizes, is important in concrete technology. The grading limits and maximum size of aggregates may affect: the relative proportions of each, cement and water requirements, workability, economy, porosity, shrinkage and durability.

From a purely economic viewpoint, it is best to use the largest size of coarse aggregate compatible with the dimensions of work. A good approximate rule is that the largest aggregate particle should not be greater than 0.2–0.25 of the smallest work dimension. For example, if the depth of a concrete floor is 150mm (6in), thus 0.25 × 150 = 38mm, practically 40mm (1.6in). This would then be the maximum size of aggregate that is permissible for this job. We may be limited by the availability of aggregate greater in size than 40mm; often 20mm (0.8in) is the maximum available, with 10mm (0.4in) also being common. Coarse aggregate is usually anything that is larger than 5mm (0.2in).

Ballast

This – also referred to as all-in ballast, all-in aggregate or combined aggregate – is a commonly used term and describes a mixture of small stones down to dust. This replaces sand plus coarse aggregate and is commonly used in DIY stores, often available in small bags. Clay and silt should not be present. The ballast must

be clean and well-graded, with around 60 per cent of particles being greater than 5mm. When using this product it is wise to make sure that it is thoroughly mixed before adding water.

Fine Aggregate

Or sand, is material that passes a 5mm sieve. Building sand is a term that may be used, along with soft sand and builders' sand. This is fine and has a small range of particle sizes; it may contain clay particles. Usual colours are white, yellow and orange. It is used 'as dug' and is not washed first. Note: this product is used for mortar and not concrete.

Sharp Sand

Sometimes called concreting sand, it has a wide range of particle sizes and feels coarse and gritty to the touch. This is used for concrete mixes and not for mortar. Silver sand comes in a wide variety of particle sizes and is used for making white concrete or white mortar (for example, for decorative purposes).

COMPACTION OF CONCRETE

It is necessary to drive out the air from a mix, that is, any air bubbles that become trapped during mixing. If any air is left in it reduces the strength and durability – as previously discussed. In order to compact the concrete thoroughly, vibration is preferred, either in the form of a beam or poker. With machine vibration it is possible to compact a lower workability concrete in the first place, thus achieving better durability. Indeed, for every 1 per cent of air left in the mix the strength and durability are reduced by 5–5.5 per cent. The practical consequences of this simple correlation are very important.

JOINTING

Joints in a concrete slab are the most likely avenue of escape for polluting efflu-ents. Therefore correct sealing is essen-tial, particularly in the construction of silage clamps. Floor slabs are subject to considerable temperature fluctuations over the course of a year, resulting in movement of the concrete. If there are insufficient joints within the slab to take account of this movement it will crack. Given a coefficient of expansion of 12 ×10⁻⁶m per degree Centigrade for typical concrete and taking an absolute extreme of 85°C variation from severe freezing to extremely hot, there would be a movement of 1mm per linear metre of concrete. In reality, a variation of 35°C is more realistic in our climate,

which equates to 0.42mm of movement per linear metre. (*Livestock Systems*, 1997)

Once the concrete has been compacted, a contraction joint is formed every 5m in the wet surface. One method, commonly used on farms, involves a 100mm (4in) wide strip of polythene laid on the surface of the wet concrete across the full width. This is pushed into the concrete with a metal bar 3mm (⅛in) thick, 50mm (2in) deep and 50mm shorter than the width of the bay. Then the bar is removed leaving the polythene-lined groove.

Every 100m in length an expansion joint must be formed to relieve the stresses that would otherwise cause cracking. They may be formed with a polyethylene board or something similar placed between the slabs to allow for expansion. The polyethylene is then routed out to make room for a sealant and performs as a bond breaker for the sealant as well. A bond breaker is essen-tial if the sealant is not to be overstressed and distorted. Expansion joints will only really be necessary when constructing farm roads.

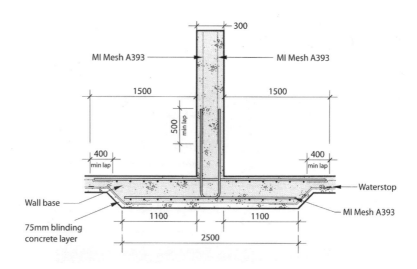

Detail of the construction of a dividing wall as built for a silage clamp. (Cemex Readymix, 2005)

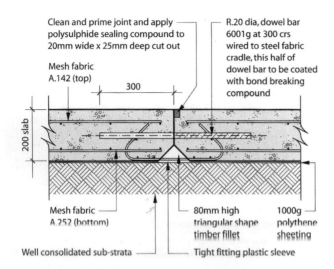

Clean and prime joint and apply polysulphide sealing compound to 20mm wide x 25mm deep cut out

Mesh fabric A.142 (top)

300

R.20 dia, dowel bar 600lg at 300 crs wired to steel fabric cradle, this half of dowel bar to be coated with bond breaking compound

200 slab

Mesh fabric A.252 (bottom)

80mm high triangular shape timber fillet

1000g polythene sheeting

Well consolidated sub-strata

Tight fitting plastic sleeve

Diagram of the expected joint requirements in concrete (in this case a paving joint detail is shown). (Cemex Readymix, 2005)

CURING OF CONCRETE

This is the process by which cement is able to hydrate. Since water is needed for the chemical reaction, we must ensure that the concrete stays in a saturated condition for a sufficient length of time to develop into a satisfactory hydrate structure. The minimum period is about a week, but this may be more with cements that hydrate slowly such as pozzolanic types.

Temperature is also an important factor – ideally, 5–15°C should be the ambient level. It should be noted that hydration is very slow at lower temperatures. Higher temperatures than these cause evaporation, together with an inferior 'gel' structure formation. Having mentioned that, concrete can be cured with care at 30°C. It is essential that concrete is never allowed to freeze before it reaches maturity; the resulting expansion of water can cause serious surface damage. When the cement has additions in it curing becomes even more important. In practice, we may cure concrete by using damp sand on the surface, hessian,

polythene sheet, curing membranes (solutions sprayed on to the surface which reduce moisture loss; the main types of membrane are resin-solvent solutions, wax (oil in water) emulsions and metallic silicates) straw or by spraying with water, for instance, by using a watering can. Full concrete strength is reached after around 28 days. As the concrete matures, it is susceptible to shrinkage and creep; these depend on the overall thickness of the section, ambient environment, time, the concrete mix and the level of water saturation. (The types of joint used for shrinkage and creep and for thermal expansion have been described above.) As an aid, to produce good concrete, the four Cs should be remembered: cement content, cover to steel reinforcement, compaction and curing. (Taylor, 1994)

TESTING OF CONCRETE

There are many ways in which concrete may be tested. This may be either when the concrete is freshly made or when it has hardened. Two common methods will be discussed here, first, the slump test – this is probably the most versatile and frequently used test and is a test of workability, secondly, the cube test will be discussed.

Slump Test (to BS 1881)

1. Empty the sampling buckets on to the mixing tray, scrape each bucket clean.
2. Thoroughly remix the sample shovelling into a heap; turn the heap over to form another, do this three times.
3. Flatten the final heap by repeatedly digging in the shovel vertically; lift the shovel clear each time.
4. Ensure the slump cone is clean and damp; place the metal plate on a solid, level base away from vibration or other disturbance; place the cone on the plate and stand on the foot-pieces.
5. Fill the cone in three equal depth layers, use the standard slump rod; rod each layer twenty-five times, spread the blows evenly over the area and make sure that the rod just penetrates the layer below; heap the concrete above the top of the cone before rodding the third layer.
6. Top up if necessary, use the rod with a sawing and rolling motion to strike the concrete level with the top of the mould.
7. Carefully clean off spillage from the sides and the base plate.
8. Carefully lift the cone straight up and clear, to a count of between 5 and 10sec.
9. Lay the rod across the upturned slump cone; measure the distance between the underside of the rod and the highest point of the concrete – the true slump; record this distance to the nearest 5mm; in all cases record the kind of slump; check the type of slump; if the

slump is not true take a new sample and repeat the test; complete the sampling and test certificate: 50mm slump = low, 75mm slump = medium, 125mm slump = high; note that high workability concrete does not necessarily have to be weaker, admixtures may be used instead. (British Cement Association, 1995)

Concrete Strength Test (to BS 1881)

Concrete test cubes are used, not to measure the in-situ strength of concrete structures, but to indicate the quality of concrete produced as a construction material for those structures. The British Standard technique for test cubes does not replicate conditions on site but establishes the quality of concrete produced. This test can be performed as outlined below.

1. Check that the moulds are clean and lightly oiled, with all bolts tightened so that there will be no leakage; ensure that the correct halves of the mould are used and that the corner lining pins are correctly located; thoroughly remix the sample as described for the slump test above.
2. Fill the mould with concrete in 50mm layers; using the special tamping bar, compact the concrete with not fewer than: twenty-five tamps for each of the two layers in a 100mm mould and thirty-five for each of the three layers in a 150mm mould; for very high workability concrete, you may not need the minimum number of tamps.
3. Remove the surplus concrete and smooth over with a float; wipe clean the mould edges; number the moulds for identification and record details.

STEP ONE
- Empty the sampling buckets onto the mixing tray
- Scrape each bucket clean

STEP TWO
- Thoroughly remix the sample shovelling into a heap
- Turn the heap over to form another
- Do this three times

STEP THREE
- Flatten the final heap by repeatedly digging-in the shovel vertically
- Lift the shovel clear each time
- If the alternative method of sampling has been used, divide this heap into two and test each part

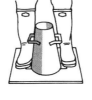

STEP FOUR
- Ensure the slump cone is clean and damp
- Place the metal plate on a solid level base away from vibration or other disturbance
- Place the cone on the plate and stand on the foot-pieces

STEP FIVE
- Fill the cone in **three** equal depth layers
- Use the standard slump rod
- Rod **each layer 25 times**
- Spread the blows evenly over the area
- Make sure the rod just penetrates the layer below
- Heap the concrete above the top of the cone before rodding the third layer

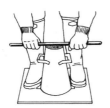

STEP SIX
- Top up if necessary
- Use the rod with a sawing and rolling motion to strike the concrete level with the top of the mould

STEP SEVEN
- Carefully clean off spillage from sides and baseplate

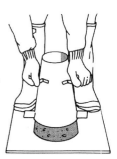

STEP EIGHT
- Carefully lift the cone straight up and clear, to a count of **between 5 and 10 seconds**

STEP NINE
- Lay the rod across the upturned slump cone
- Measure the distance between the underside of the rod and the **highest point** of the concrete - the true slump
- Record this distance to the nearest 5 mm
- In all cases record the kind of slump
- Check the type of slump
- If the slump isn't true, take a new sample and repeat the test
- If the second slump isn't true, get advice
- Complete the *Sampling and testing certificate*

There are three kinds of slump

1. TRUE

Slump

2. COLLAPSE

3. SHEAR

The slump test (to BS 1881: Part 102) is the accepted technique of testing fresh concrete for its workability. (BCA, 1995)

4. Cover each mould with a damp cloth and plastic sheet; store inside at normal room temperature (15–25°C), for instance, on top of the curing tank; protect the cube moulds at all times from high and low temperatures, especially frost and high winds; complete the sampling and cube making certificates.

The cubes should be taken from the moulds the day after they were made (24hr later), numbered and put into a curing tank. The sides of the moulds are tapped gently with a hide hammer and then lifted off. The cubes are marked in order to identify them and then they are put into the tank. The water temperature of this curing tank must be controlled at 20°±2°C.

One week after the cubes were made they should be removed from the water tank and crushed in order to ascertain

Sample the concrete in the standard manner. Concrete sampled using the alternative method cannot be used for making cubes.

STEP ONE

- Check that the moulds are clean and lightly oiled with all bolts tightened so that there will be no leakage
- Ensure that the correct halves of the mould are used and that the corner lining pins are correctly located
- Thoroughly remix the sample as described for the slump test and flow test

STEP TWO

- Fill the mould with concrete in 50 mm layers
- Using the special tamping bar, compact the concrete with **not less than:**
   25 tamps for each of the two layers in a 100 mm mould
   35 tamps for each of the three layers in a 150 mm mould

For very high workability concrete you may not need the minimum number of tamps.

STEP THREE

- Remove surplus concrete and smooth over with a float
- Wipe clean the mould edges
- Number the moulds for identification and record details

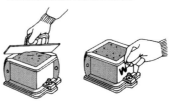

STEP FOUR

- Cover each mould with a damp cloth and plastic sheet
- Store inside at normal room temperature (15°C to 25°C) e.g. on top of the curing tank
- Protect the cube moulds at all times from high and low temperatures (especially frost) and drying winds
- Complete the *Sampling and cube making certificates*

The cube test (to BS 1881: Part 108) is an accepted method of testing the strength of concrete. The diagram shows the stages in making the cubes correctly. (BCA, 1995)

STEP ONE
- Record the maximum/minimum overnight storage temperatures on the certificate

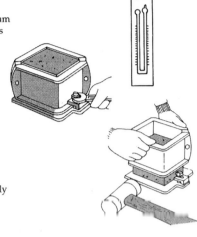

STEP TWO
- Slacken off all nuts

STEP THREE
- Part the sides of the mould, tapping gently with the hide hammer
- Lift off carefully
- Remember, new cubes are easily damaged unless handled carefully

STEP FOUR
- Mark each cube with its identification number on two of its cast sides

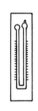

STEP FIVE
- Place the cubes in the curing tank
- Clean and reassemble the moulds

STEP SIX
- Check that the water temperature is controlled at 20°C ± 2°C and the cubes are covered by water
- Make sure the power supply is not switched off day or night
- Check the temperature range daily using the maximum/ mimimum thermometer
- Keep a record of the readings

STEP SEVEN
- For despatch to test labratory, wrap the wet cubes in damp cloths, and then plastic bags and pack in trays
- Attach the sampling, testing, cube making and storage certificates, plus the order for testing, to the package

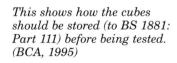

This shows how the cubes should be stored (to BS 1881: Part 111) before being tested. (BCA, 1995)

their 28-day strength. Normally, three cubes of each strength being tested will be made and crushed and the highest reading for the three is taken, to make allowance for a poor result in the experimentation. The results for this test will show the quality of the concrete produced, and, if different strengths are compared in this way, an interesting set of results will be gained.

THE SAFE USE OF CEMENT

Health Hazard

Dry cement powders in normal use have no harmful effect on dry skin. As with any dusty material, there may be ill effects from the inhalation or ingestion of cement dust and suitable precautions should be taken. When cement is mixed with water, alkali is released and precautions should therefore be taken to prevent dry powder entering the eyes, mouth or nose and to avoid skin contact with wet concrete and mortar. Repeated skin contact with wet cement over a period may cause irritant contact dermatitis and the abrasiveness of the mortar or concrete constituents can aggravate the effect. Some skins are sensitive to the small amounts of chromate which may be present in cements and can develop allergic contact dermatitis, but this is rare. Continued contact with the skin can result in 'cement burns' with ulceration.

Handling Precautions

Protection for the eyes, mouth and nose should be worn in circumstances when dry cement may become airborne. When working with wet concrete or mortar, suitable protective clothing should be worn, such as long-sleeved shirts, full length trousers, waterproof gloves with cotton liners and Wellington boots. Clothing contaminated with wet cement, mortar or concrete should be removed and washed before further use. Should concrete or mortar get into Wellington boots, remove them immediately and thoroughly wash the skin and the inside of the boots before proceeding with the job. If cement enters the eye it should be washed out immediately and thoroughly with clean water and medical advice sought. Concrete or mortar elsewhere on the skin should also be washed off immediately. Whenever there is persistent or severe irritation or pain, a doctor should be consulted.

Spillage

Collect or remove for reuse. Suitable respiratory protection equipment should be worn to protect against airborne dust. Care should be taken to avoid contamination of cement recovered after spillage.

First Aid

Skin contact – wash with soap and water immediately; if there is irritation or pain, seek medical advice. Eye contact – wash with plenty of clean water and seek medical advice. (British Cement Association, 1991)

ABOVE: A converted farm building in a diversification enterprise. In the foreground is a 'farm park museum'.

LEFT: Another view of the same building. Old farm machinery can be seen in the foreground.

BELOW: The same farm; redundant farm buildings have been turned into a museum.

CHAPTER 5

Foundations, Walls, Floors, Roof and Frame

The building elements have to withstand the following loads: snow, wind, dead and imposed. The most important dead loads will include the self-weight of the building structure; the imposed load will consist of both dead and live loads, live loads will include people and livestock and dead load comprises furniture, moveable partitions and other equipment. Both the wind load and the snow load should be self-evident and will be determined by geography primarily. They utilize fairly complicated formulae (and these days computer programs), particularly the wind load and hence will not be further discussed here.

There are many factors that will affect the choice of materials used to form a building. These may include:

- length and height of the walls
- the loads that will be carried
- degree of fire resistance that will be required
- whether roofs or floors are to be clear span
- type of roof structure
- aesthetics
- weatherproofing of the components.

The substructure of a building is that situated below the damp-proof course (DPC), and its function is to transfer the imposed and the dead load to the ground.

Important parts of the substructure therefore are the foundations and the floor.

FOUNDATIONS

The function of the foundations is to 'carry the loads of the building and distribute them over the ground in such a way that movement of the building is minimal'. (Millward, D., 2000) The foundation will be constructed using concrete, maybe reinforced, and will fit into one of the following types: strip, pad, raft or pile. The choice of foundation will depend on the conditions of the site and the soil present; the table overleaf, top, shows the types of foundation that might be applicable.

In order for the substructure to perform satisfactorily when in use it must transmit the dead and the imposed load from the building down into the ground, so that total movement is limited; it should avoid damage from shrinkage, swelling or freezing of the subsoil and it must resist attack by sulphates or other materials in the soil. To calculate the size of foundations necessary in certain soil types the table overleaf, bottom, may be used, this gives allowable bearing pressures.

Possible Types of Foundation for Various Soil and Site conditions

Soil and Site Conditions	Possible Type of Foundation
Rock or solid chalk, sand and gravels	Shallow strips or pads
Firm stiff clay with little vegetation liable to cause shrinkage or swelling	Strips 1m (3ft) below ground level or piles with ground beam
Firm, stiff clay with trees close to the building	Trench fill or piles and ground beam
Firm, stiff clays where trees have recently been felled and the ground is still absorbing moisture	Reinforced piles or thin reinforced rafts in conjunction with a flexible building structure
Soft clays or soft silty clays	Wide strips; up to 1m wide or rafts
Peat or sites consisting partly of imported soil	Piles driven down to a firm strata of subsoil
Where subsidence might be expected (for instance, mining districts)	Thin reinforced rafts

(Millward, D., 2000)

Bearing Capacities of Various Types of Ground

Classification	Bearing Capacity (kN/sq.m)
Rocks	
strong sandstone	4000
schist	3000
strong shale	2000
Granular soils	
dense sand and gravel	>600
medium dense gravel	200–600
loose sand and gravel	<200
compact sand	>300
loose sand	<100
Cohesive soils	
stiff boulder clay	300–600
stiff clay	150–300
firm clay	75–150
soft clay and silt	<75

(Millward, D., 2000)

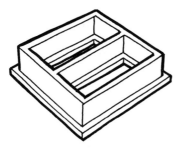

A strip foundation that may be suitable for many buildings. Each wall is supported by a continuous strip of concrete.

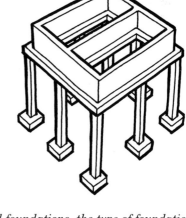

Pad foundations, the type of foundation often used under farm buildings.

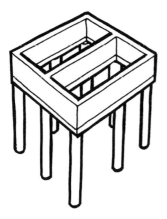

Pile foundations. An alternative type of foundation where the subsoil immediately beneath the building is weak.

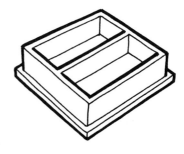

Raft foundations; this design would be suited for unstable subsoil. A continuous concrete raft supports the weight of the building.

The main types of foundation are illustrated above.

In order for the foundation to be durable, the concrete must be of the correct type. The density of the concrete will normally be 2150–2500kg/cu.m with a 28-day compressive strength (measured by the cube test) of 15–60N/sq mm. The water : cement ratio is likely to be 0.4–0.7 for the foundation mix. Note that if the soil contains a sulphate that could lead to deterioration of the foundations, sulphate-resisting cement should be used.

There are a few simple rules that are used when designing farm buildings to determine how thick and how wide foundations should be and how deep they should go into the ground. Generally speaking, the foundations must be wide enough to spread the imposed load over a large enough area so that settlement or movement will not cause damage to the building nor threaten stability. The footing must also be deep enough to safeguard the building against damage by swelling, shrinkage or freezing of the sub-

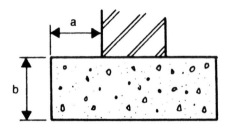

Diagram of a typical strip foundation. Here the distance from the outer sides of the wall (a) must never be greater than the depth of the footing (b). (Marley Eternit, 1991)

may well design the foundations himself, and in this case the farmer will need to obtain the relevant details and specification from the manufacturer; see the figure below for typical footing design (Eternit, 1991).

With buildings where heavy machinery will be driven in and out, such as a grain store, it is wise to provide columns with a wheel guard in the form of a concrete kerb cast around the foot of the column. The kerb should extend from the foundation level to whatever height is needed in view of the size of the machinery involved in order to protect the column.

soil. With strip foundations for walls, the distance from the outer sides of the wall must never be greater than the depth of the footing (see the figure above).

In portal frame buildings, the weight of the roof is transmitted to the ground by the supporting columns which rest on concrete foundations. The manufacturer

FLOORS

The type of floor required varies a great deal and depends upon the function of the building. Some buildings such as cattle yards require a free-draining floor to keep the bedding as dry as possible, some stock

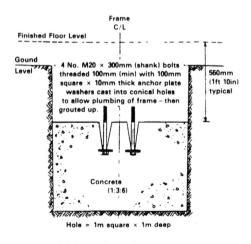

(a) Portal frame foundation

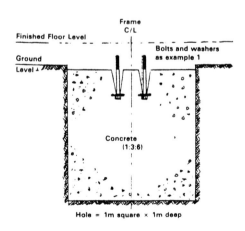

(b) Portal frame foundation suitable for a relocatable building

Diagrams showing typical footing design suited for agricultural buildings. (Marley Eternit, 1991)

require insulated floors, almost any hard dry surface will suffice for farm machinery, while grain stores must be absolutely dry and damp-proof. Earth floors require little description except that they are suitable only on well drained sites. Floors of hardcore, gravel, clinkers, chalk and hoggin are all useful in their proper place, as are those of pulverized fuel ash from power stations and colliery shale, but only in cases where sulphates are known to be absent.

One ready-mix concrete company, RMC Readymix, produces a range of four purpose-designed agricultural mixes:

- Readymix Farmpave – designed for use on external paved areas subject to the constant loading and scraping imposed by farm machinery and vehicles; it is designed to resist damage from frost and aggressive conditions found in an agricultural environment.
- Readymix Liquitite – specially developed for use in silage clamp and waste slurry design; it is a high-strength, durable concrete designed to meet the stringent conditions and legislation for this type of design.
- Readymix Multistore – a durable and load-bearing concrete designed for floors in both general and specific storage facilities; with a good structural design and the use of this product a hygienic storage facility can be created.
- Readymix Stockfloor – a high quality concrete that has been designed to satisfy the standards of hygiene required in today's livestock housing including the Food Safety Act and Quality Assurance Schemes (RMC Readymix, 2005).

Readymix Farmpave will now be described in more detail; it is designed to be:

- strong – withstands the imposed loads and resists the damage from modern farm machinery; it is also designed to reduce the risk of plastic cracking, preventing excessive weathering;
- durable – a high-strength mix giving low porosity concrete which will reduce the water absorption in the top surface allowing it to maintain the wearing course which in addition aids against frost damage;
- practical – in farm building design, floors and surfaces are rarely laid flat or level and concrete can be formed in order to assist drainage and surface water control, it also creates a cleanable surface; this is essential for compliance with modern food standards;
- assured – the mix is designed to meet the requirements of BS 5502 (Part 21).

Increasingly within 'quality assurance' systems external yards are viewed as important as the production facility or the enterprise itself with reference to yard roads being specified in QA standards, for example, Milk and Dairy Regulations and National Dairy Assurance. In order to ensure that the concrete not only passes inspection but is also durable and strong, it is essential to place and cure it properly.

Pre-construction

Before beginning the project it is important to carry out some obvious checks and measurements:

- check for lorry access
- assess the ground conditions, noting made-up or soft ground
- ensure that adequate resources are provided in the form of people and equipment to place, compact and finish the concrete

- ensure that the sub-base and forms are properly prepared (poorly compacted sub-base or poorly fixed forms often result in concrete failure)
- plan ahead for extreme temperature, both high and low, since this can seriously affect the final strength of the concrete; in low temperature cover with an insulating material and in high temperatures cover with polythene and spray with water.

Construction

- Base works: excavate and fill with graded granular material compacted and raised in 225mm (9in) lifts up to sub-base level. For the sub-base lay graded granular material, and then blind it with a thin layer of sand and compact with a vibrating roller.
- Reinforcement: this mix has been developed to eliminate the need to provide crack control steel reinforcement (top steel). If there is no anticipated

A roller used for compacting the sub-base before concreting.

ground movement then no reinforcement is required.
- Finishing: texture the concrete to provide a finished surface. A heavy brush or tamped finish is the most common. You will also need to backfill the edges of the concrete with sub-base material to avoid the cracking of the edge and to ensure that rain can run off the low side of the concreted area.

Post-construction

Check periodically for deterioration of joints and replace jointing material if required. Failure to seal the joints will result in the onset of the freeze/thaw process in the joint.

Readymix Agricultural Products – Floor Finishes

The table on p.32 shows the mixes that Readymix supplies, together with the recommended finish, the agricultural requirements, relevant quality assurance and regulations, benefits and some notes.

WALLS

Often, walls used in portal framed buildings are dwarf walls constructed from brick or precast concrete blocks up to a height of around 1.5 to 2m (4.5–6ft). Alternatively, walls can be built using large sections of precast concrete, often equal to the bay width of the building. These may offer a speedy type of cladding that may be suitable for stock or crop loadings. Sometimes railway sleepers are used, not as common nowadays, or else the cladding may extend down to ground

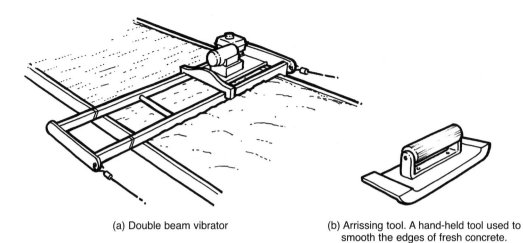

(a) Double beam vibrator

(b) Arrissing tool. A hand-held tool used to smooth the edges of fresh concrete.

Tools to aid the compaction of fresh concrete.

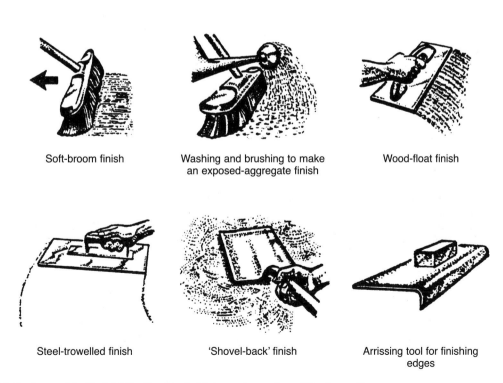

Soft-broom finish

Washing and brushing to make an exposed-aggregate finish

Wood-float finish

Steel-trowelled finish

'Shovel-back' finish

Arrissing tool for finishing edges

Methods used to 'finish' the freshly laid concrete surface. The type of surface developed will depend on its proposed uses.

level where there is no direct loading to be put on the walls. Concrete block or brick dwarf walls are popular because of resistance to accidental damage from livestock and machinery, and they also offer advantages when it comes to planning permission and the overall 'view' of the farm buildings. Again, depending on the requirements of the building, fibre cement sheeting, space boarding or other types of cladding can then be used above the dwarf walls.

A wide range of concrete blocks is available and quality varies more than cost; the priority for the farmer is to choose the correct type in relation to the function of the building. Lightweight blocks, for instance, offer good insulation and therefore could be the ideal material where the retention of heat (or cold) is important, as in a cold store. However, where strength is the most important factor dense blocks are better.

Another factor to bear in mind as well, as providing sound foundations, is the correct mortar to use. The mortar used for block laying should be weaker than the blocks so that any cracks caused by wall movement or block shrinkage will occur at joints where they will not be as unsightly. They may then be refilled more easily than having to attempt to replace wall blocks or bricks. A cement-lime-sand mix (with builders' lime) or a mix of masonry cement and sand should be used. For external work the recommended proportions for a cement-lime-sand mix are 1 : 1 : 6, except in exposed positions or for block laying in winter; in these cases the strength of the mix should be increased to 1 : 0.5 : 4.5. There are several proprietary admixtures available which make the mortar more workable and eliminate the use of lime. These have

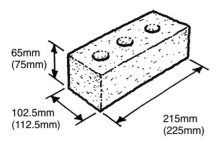

Sketch of a typical standard brick showing dimensions without and (in brackets) with mortar.

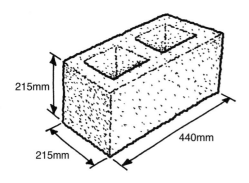

A hollow dense block equivalent to
six standard bricks

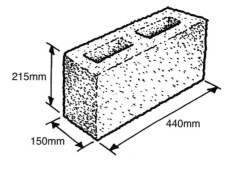

A hollow, cellular dense block

Diagram with the dimensions of concrete blocks as used in farm building construction.

been described in Chapter 4. When the mix is right, the mortar should stick to the vertical sides of the blocks without any tendency to fall away. See the accompanying figures, showing typical brick and block dimensions and a technique of block laying.

Laying Blocks

The arrangement of the blocks and the size of the building should be carefully worked out before work begins so that the foundation level, the floor level and the sizes of door and window openings can be fitted in with the coursing and the lengths of standard blocks. Lintels should be of the same depth as one course of the block work and can be built up from special lintel or bond beam blocks. Block cutting should be avoided as much as possible; the use of 0.5-size and 0.75-size blocks helps to reduce this. If cutting is necessary, then the best tool to use for dense blocks is a bolster. Lightweight concrete blocks can be sawn by using an ordinary handsaw.

Block laying is done in much the same way as bricklaying. The courses must be laid straight and level and each plumbed up with the last. The blocks must be bonded, that is, each block laid so it overlaps the one below by half its length. Block walls should not be built to a height of more than 1.2m (4ft) at one time. The mortar should then be allowed to harden for one or two days, depending on the temperature, before more blocks are laid. Long stretches of block work unsupported by cross walls must be strengthened and given stability by means of piers made up of concrete blocks. The piers should be built at intervals depending on the thickness and the height of the wall. In framed buildings

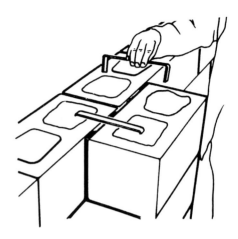

Note that the pier is built alongside the wall and is not bonded into it. The pier and the wall are tied together by metal straps. The concrete infilling extends for the full height of the wall and pier, thereby providing extra strength.

Sketch showing how a hollow concrete wall pier should be built. (Marley Eternit, 1991)

columns with metal ties projecting into the block work help to perform this function. If the wall is not to be rendered, the joints should be pointed; the object of this is to compress the mortar into a dense, compact mass and, at the same time, to give the joints a good appearance. Pointing should be shaped to shed water. If the wall is to be rendered, the joints should be raked out to a depth of about 12mm (0.5in) to provide a good 'key' for the rendering. Newly constructed block work should be covered to protect it from the effects of rain or frost. A plastic sheet or tarpaulin will prevent the blocks from being soaked by rain and sacks will provide protection against frost. If the blocks are not kept dry during construction their drying out later may cause unnecessary shrinkage cracks in the finished wall.

Damp-proof Course

If a damp-proof course is necessary, it should be placed near the base to prevent ground moisture from rising up the wall. It should be placed just high enough to be above any splashing by heavy rain, and normally not less than 150mm (6in) above the ground. If the floor inside the building is below the level of the ground outside, a vertical damp-proof course is also required. Polyethylene and lead-cored bituminous felt are the two most popular types of damp-proof course. They must be laid on a solid, smooth surface. In hollow block work therefore the core holes must be filled with concrete up to the level of the damp-proof course. At joints in the strips there should be a good overlap to ensure that leakage cannot occur.

Single-thickness Walls

Walls one block thick of hollow, cellular or solid blocks are adequate for most farm construction jobs. If hollow blocks are used, weight and strength may be added by filling the holes with fine concrete. This should be done a few courses at a time, and the concrete should be well rammed. Hollow block work may also be reinforced by inserting steel rods vertically into the core holes of the blocks and then packing the holes firmly with concrete. This is particularly useful at points where lateral loads may be imposed on the wall or where the wall needs strengthening, as at door jambs. In general, these bars should have a minimum concrete cover between themselves and the surrounding block of 20mm (0.8in). The bars may be placed in 1.2 or 1.5m (4 or 5ft) lengths, provided that each new length is lapped to a minimum distance of 300mm (1ft) along the previously placed bar and wired to it.

Before placing the reinforcement in position, all loose rust and scale should be removed from the bars by brushing with a stiff wire brush. If an exceptionally strong wall is needed, steel reinforcing rods may be set in the horizontal joints. To bond a hollow block wall firmly to its base, L-shaped 'starter bars' must be cast into the foundation at about 0.5m (1.5ft) intervals to coincide with the hollows in the blocks.

Although single leaf walls are never quite impervious to moisture, walls of hollow blocks which have been 'shell'-bedded are usually more resistant than walls of solid blocks. External cement rendering will improve moisture resistance, as will two coats of cement paint. In many farm buildings slight moisture penetration will not be of great consequence, but, if an absolutely impervious wall is required, cavity construction should be considered or a single leaf fully weathered externally (Marley Eternit, 1991).

ROOF

Typically, the roof of a farm building may be covered with one of two materials, either steel or fibre-cement sheeting. There are one or two other alternatives which will be explained below.

Fibre-cement

Most fibre-cement products may be guaranteed for periods of 30 years (in fact, roofs covered with the old asbestos cement were deemed to have a life of 40 years; asbestos has now been replaced on safety grounds, see Chapter 15). The

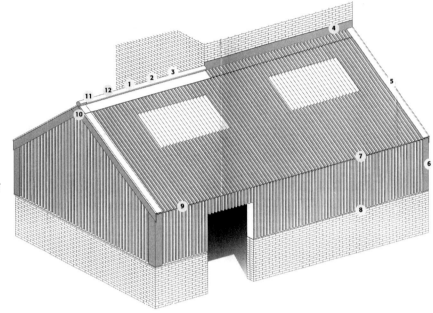

Typical roofing details as depicted by Eternit. The numbers on the diagram are referred to in the figure below. (Marley Eternit, 2004)

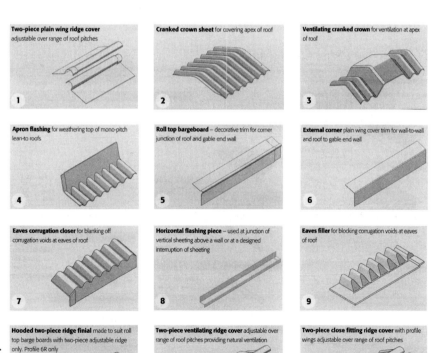

1 Two-piece plain wing ridge cover adjustable over range of roof pitches

2 Cranked crown sheet for covering apex of roof

3 Ventilating cranked crown for ventilation at apex of roof

4 Apron flashing for weathering top of mono-pitch lean-to roofs

5 Roll top bargeboard – decorative trim for corner junction of roof and gable end wall

6 External corner plain wing cover trim for wall-to-wall and roof to gable end wall

7 Eaves corrugation closer for blanking off corrugation voids at eaves of roof

8 Horizontal flashing piece – used at junction of vertical sheeting above a wall or at a designed interruption of sheeting

9 Eaves filler for blocking corrugation voids at eaves of roof

10 Hooded two-piece ridge finial made to suit roll top barge boards with two-piece adjustable ridge only. Profile 6R only

11 Two-piece ventilating ridge cover adjustable over range of roof pitches providing natural ventilation

12 Two-piece close fitting ridge cover with profile wings adjustable over range of roof pitches

The several roof components according to Eternit. (Marley Eternit, 2004)

Mitring plan: single slope roof

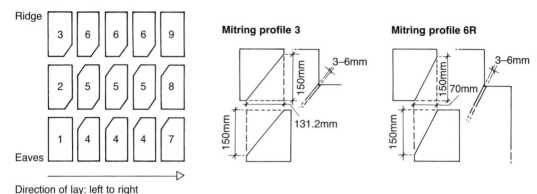

Direction of lay: left to right
Mitres opposite hand for laying right to left

Details of roof sheets as they are fixed to a farm building. (Marley Eternit, 2004)

advantages of using a fibre-cement roof are seen to be:

- it is a cementitious product with no metallic content, therefore there is no risk of corrosion
- it is suitable for use in aggressive or marine atmospheres
- it is minimally affected by frost or climatic temperature changes
- its vapour permeability significantly reduces the risk of condensation; the sheeting is able to absorb up to 25 per cent of its dry weight in moisture and dissipate it in more favourable conditions
- it has low thermal conductivity that reduces the heat build-up in summer and heat loss in winter
- it provides good acoustic insulation, with substantial mean airborne sound reductions
- there is a good range of colour finishes.

Eternit produce two ranges of profiled sheeting, Profile 3 and Profile 6R. The latter is a fibre-cement sheet with polypropylene strips inserted along precisely engineered locations which run for the full length of the sheet in each corrugation. This provides maximum reinforcement strength with no loss of durability.

Fixing

- the sheets should be installed smooth surface up
- the sheets should be cut with a hand-saw or slow-speed, reciprocating power saw
- all fixing holes should be drilled, not punched, and should provide adequate clearance for the fastener shank (minimum 2mm)
- there should be two fixings per purlin or rail covered at the fixing points
- when using power tools in a confined area, dust extraction equipment is advisable
- the dust and swarf generated when working with the sheets does not require any special handling requirements other than normal good housekeeping to maintain a clean working area.

The correct fixing of a sheet is important in order to avoid premature failure, corrosion or leaks in a roof. Many factors influence the fixing of a roof, such as the purlin or rail type and the nature of the roof in question; particularly important is the type of fastening system used and compliance with the manufacturer's recommendations.

For the fixing of Eternit Profile 6R sheets on roof slopes up to 30 degrees the recommended method is the use of topfix fasteners. These provide a quick and effective, one-step fixing operation. However, they must be installed by using the recommended depth-locating power-tool to prevent under or over tightening, which can damage the roof sheets. In all instances, the sealing washers and pro-tective caps should be utilized to ensure adequate weather protection.

Overhangs

Sufficient overhangs must be allowed at the eaves to ensure that rainwater discharges into the gutter. Verges must be overhung by one complete corrugation unless a bargeboard is used.

Side and End Laps

Where appropriate, 8mm-diameter butyl strips should be positioned, as shown in a preceding figure. The minimum end lap for either Profile 3 or Profile 6R is

OVERHANGS

Sufficient overhangs must be allowed at the eaves to ensure that rainwater discharges into the gutter.

Verges must be overhung by one complete corrugation unless a bargeboard is used.

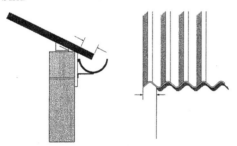

END LAPS

The minimum end lap for either Profile 3 or Profile 6R is 150 mm, fixed as shown in the section below.

Where double sealing is necessary, the end lap should be 300mm and the second butyl strip should be positioned 100–200 mm below the fixing.

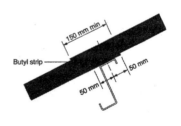

SIDE LAPS
Sealing
Where appropriate, 8 mm diameter butyl strips should be positioned as shown.

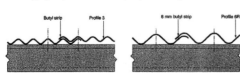

CHECKING THE TOPFIX FASTENERS FOR TIGHTNESS

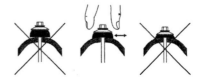

The stages of roof fixing, after Eternit. Note that most farm buildings now use corrugated fibre-cement materials. (Marley Eternit, 2004

150mm (6in), fixed as shown in the figure on p.59. Where double sealing is necessary the end lap should be 300mm (1ft) and the second butyl strip should be positioned 100–200mm (4–8in) below the fixing.

Fixing Detail

The fixing of a fibre-cement roof can be accomplished by most people if they follow these ten steps. In order to weatherproof the roof the butyl strip must be installed as described and mitres cut to avoid having four thicknesses of sheeting in the same plane at the junctions of sides and end laps.

1. lay sheet #1 at the eaves without mitring
2. lay sheet #2, mitring bottom right hand corner as shown in the diagram
3. lay sheet #3, mitring as before; continue up roof slope to complete the first tier
4. lay sheet #4 at the eaves of the next tier, mitring the top left-hand corner as shown
5. lay sheet #5, mitring both top left-hand and bottom right-hand corner as in the illustration; continue up the slope until ready to lay sheet #6
6. lay sheet #6 at the ridge, mitred as in step 2
7. repeat the procedure from and including step 4, working across the roof from eaves to ridge, until there is room for one more tier to be laid on the right-hand edge
8. lay sheet #7, mitring the top left-hand corner; if necessary, reducing the sheet width by cutting down the right-hand edge; all subsequent sheets in this final tier should be cut accordingly

9. lay sheet #8 as in step 7, continuing up roof slope until ready to lay the final sheet at the ridge
10. lay sheet #9 at the ridge without mitring to complete the roof.

It is important to note:

1. on a duopitch roof start both slopes from the same end of the building; one slope will therefore be sheeted left to right, the opposite slope will be sheeted right to left
2. the corrugations of sheets must line up at the apex to ensure that the ridge accessories will fit
3. when cranked crown sheets are used, both top courses of roofing sheets and the cranked crowns themselves must be mitred
4. always lay sheets with the correct end and side laps
5. do not cut mitres *in situ*.
(Marley Eternit, 2004)

Steel Sheeting

In order to fix steel sheets properly it is wise to lay the first one at the eaves, at the end of the building furthest away from the prevailing wind, to reduce the potential for rain to be driven in at the side laps. Normally, ridge to eaves would be covered with one sheet, but where two courses are necessary, the first course should be laid along the eaves (see accompanying figures).

It is essential that the first sheets of each course are set square to the building frame and that, as work progresses, a check is made as each building frame is passed. Unless frequent checks are made on the alignment, the roof will be out of line at the gutter or the ridge or both and at the gable ends. This results in an

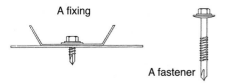

A fixing

A fastener

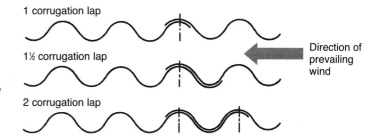

Typical primary fasteners

Self-tapping | Self-drill/ self-tapping | For timber

For steel

Primary fastener

Secondary fastener

'Primary fasteners and fixings' – attach cladding to the building framework.
'Secondary fasteners and fixings' – attach sheets to one another or attach flashings and other accessories to sheets.

Sheet steel details, including fastenings and fixings required.
(British Steel Strip Products, 1992)

1 corrugation lap

1½ corrugation lap

2 corrugation lap

Direction of prevailing wind

RIGHT: *Diagram to show how adjacent roof sheets (steel) should overlap. (British Steel Strip Products, 1992)*

unsightly result and also in rainwater overshooting the gutters, the ridge caps not fitting snugly and the gable flashings and bargeboards being out of line and difficult to fix.

Galvanized steel purlins and cladding rails are recommended; treated timber purlins and rails must be isolated from the sheeting by a layer of DPC-type material to prevent the salts in the treated timber from corroding the sheet. When fixing box profiles it is recommended that fasteners are located in the valleys of the profile. Long sheets that cover the full length from the ridge to the eaves should be used whenever possible; this avoids end laps and thereby precludes any possibility of leaking or corrosion at this point. If end laps are required, laps of 150mm (6in) are satisfactory in most situations. In exposed sites end laps should be increased to

250mm (10in). On roof pitches less than 10 degrees end laps should be sealed with a flexible sealant. With box (trapezoidal) profiles the amount of side lap and the fastener details will be determined by the particular profile being used. With standard corrugated (sinusoidal) steel profiles side laps of one-and-a-half corrugations are usually satisfactory; in exposed conditions it is advisable to increase the side laps to two corrugations. In sheltered locations, one corrugation of overlap will be adequate. Side laps should be secured with secondary fasteners that are not more than 450mm (18in) apart.

The following are some recommendations on using fasteners with steel sheeting:

• holes through sheeting are best drilled, not punched

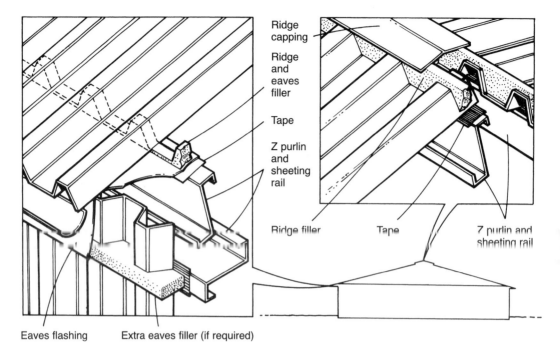

Ridge capping

Ridge and eaves filler

Tape

Z purlin and sheeting rail

Ridge filler   Tape   Z purlin and sheeting rail

Eaves flashing   Extra eaves filler (if required)

Constructional details of a steel roofed barn.

- the hole size should be drilled to match the type of fastener being used; it is preferable to use self-drilling, self-tapping fasteners since these ensure that the holes are of the correct size
- fasteners should be perpendicular to the sheet surface to ensure a good seal by the fastener washer
- fasteners should not be over-tightened otherwise the washer will deform and not seal properly; special tools can be hired which will ensure the correct alignment and tension of fasteners
- consult the manufacturer to ensure that the correct type of fastener is being used, particularly regarding its degree of corrosion-resistance; note: the fasteners for fixing to metal will differ from those for fixing to timber
- timber purlins and rails should be of at least S3 grade, with a moisture content not greater than 25 per cent, otherwise, when the timber dries out, the fixing will be less secure; fixings in timber should be at least 25mm (1in) from the edge of the timber; otherwise the timber may split and weaken the fixing
- timber should not be treated with preservatives containing copper, mercury or other metal salts as these will corrode sheets and fasteners; if treated timber is used, a DPC-type material should separate the timber from the sheeting (British Steel Strip Products, 1992).

Insulation

Since steel sheeting offers little in the way of insulation, it may be specified for use with some form of it. This will depend on what the building is to be used, as, of course, with cold stores being an example where insulation would be used. Rigid insulation board can be fixed over, between, or under purlins; the other method is to use composite panels.

The diagram opposite shows details of the insulation that is required when using steel sheeting.

'Polytunnels'

Polytunnels used in agriculture may offer certain advantages:

- they are cool, dry, light and draught-free
- they are a clear span structure offering a completely open space
- available in spans from approximately 6 to 24m (6.5–26yd)
- they may have opaque roof coverings combined with netted sides, to produce

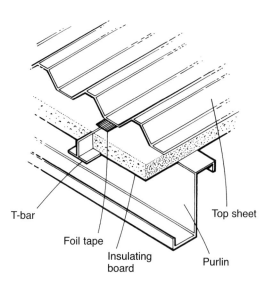

T-bar | Top sheet
Foil tape
Insulating board | Purlin

Steel sheet used in conjunction with insulation to reduce the problem of cold-bridging. This type of construction is often used with insulated crop stores.

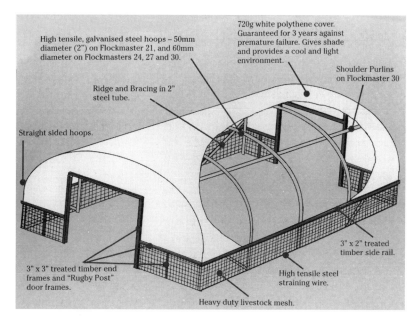

High tensile, galvanised steel hoops – 50mm diameter (2") on Flockmaster 21, and 60mm diameter on Flockmasters 24, 27 and 30.

720g white polythene cover. Guaranteed for 3 years against premature failure. Gives shade and provides a cool and light environment.

Shoulder Purlins on Flockmaster 30

Ridge and Bracing in 2" steel tube.

Straight sided hoops.

3" x 2" treated timber side rail.

3" x 3" treated timber end frames and "Rugby Post" door frames.

High tensile steel straining wire.

Heavy duty livestock mesh.

Layout of a plastic-clad tunnel, as used for sheep housing. (National Poly-tunnels, 1997)

a shaded and well-ventilated environment (they may be fitted with optional ridge/side vents)

- often available in several colours to comply with planning restrictions
- the tubes are usually galvanized steel, thereby offering good corrosion resistance
- the ground inside may be left as bare soil which will bake rock-hard to form an ideal, free-draining base for sheep and make mucking out easy
- there are several types of covering that may be used, for instance, white polythene (with a three-year guarantee), Duraweave fabric (from 'Cover-all' with a 15-year guarantee).

This type of agricultural building may offer certain advantages to the farmer, including having a shelter that is quick to erect and may be relatively cheap (depending on the design). It may also be possible to move the building at a later date if so desired (National Polytunnels Ltd, 1997). The figure on p.63 shows the constructional features of a plastic tunnel.

FRAME

The frame for a farm building will be offered as:

- concrete – normally the most expensive option but should be the most durable; it is generally the most difficult to fix into; such frames are now difficult to buy, certainly as new, and often do not blend as well from an aesthetic point of view; they are generally a building erected in the second half of the last century
- timber – usually more expensive than

steel but generally more aesthetically pleasing; it is certainly more likely to gain planning permission if in a difficult situation and because it is wood it will not last as long as the other two options

- steel – this is the least expensive and therefore the most common; it does have higher maintenance costs and requires special protection in some circumstances, for instance, when used for a silage store; it is the most flexible, permitting a greater range of design in terms of bay size, span, height, length and roof pitch.

The next figure shows the main component parts of typical farm building and the following three show one manufacturer's designs of farm building.

Spans

The spans of steel or concrete buildings may be single or multi-span. The width of span, particularly for steel, may be any size up to around 30m (33yd) or more. Spans commonly used in agriculture are 9m up to 25m (10–27yd). A lean-to is often found in farming situations and may be full or variable pitch; the span may be limited by the height to the eaves of the main building and the eaves height required in the lean-to itself. Spans commonly found in agriculture are 4.5 to 12.5m (5–13.6yd).

Bay Length

Commonly, these are 4.8m and 6m (5.3–6.5yd), but there are other sizes available. The bay spacing is measured from centre to centre of the stanchions and the total length of the building

Essential features of portal frames

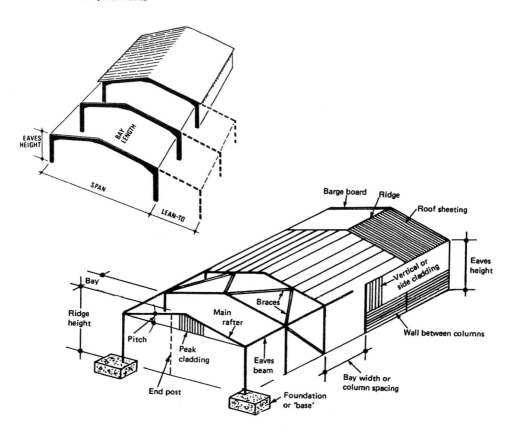

Traditional truss

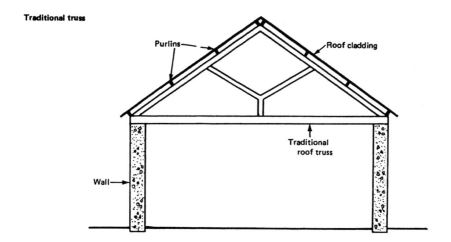

Three figures to depict the main names of traditional farm building parts. (Noton, 1982)

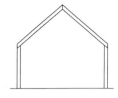

Monopitch
Up to 8m width clear span. Over 8m width with intermediate posts. Optional cantilever either straight or drop snout.

Supported apex
Up to 12m width. Most suited to cattle or sheep housing, with varying passage widths.

Trussed type clear span
Varying widths available up to 12m span.

Sussex type barn
Constructed to a traditional style up to 10m wide.

One manufacturer's building profiles as available. (Farmplus Construction Ltd, 2005)

Single bank kennel unit

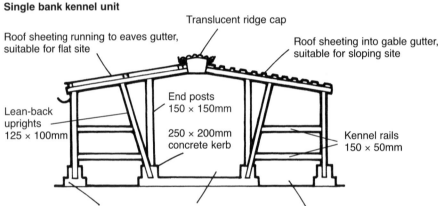

House and feed

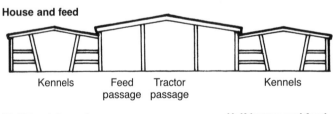

Multi-bank kennels

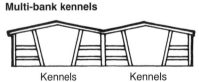

Half-house and feed

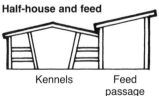

The layout of one manufacturer's kennel unit, with main dimensions, for cattle. (Farmplus Construction Ltd, 2005)

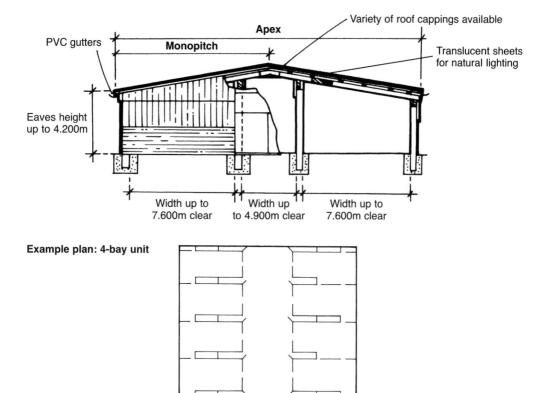

A timber building, with dimensions, as used for a cattle shed. (Farmplus Construction Ltd, 2005)

is taken to be the sum of the bay spacings.

Height

Of course, this will vary according to the requirements of the building, with a maximum usually of 6m (6.5yd). Over recent years the height of buildings has increased to cater for the needs of, primarily, machinery. Eaves height is measured from the finished floor level to the upper side of the eaves beam, in other words, to the underside of the cladding material.

Roof Pitch

This will be variable depending on air space requirements and on planning stipulations (since the slope will have a large bearing on the building's visibility from a distance). Pitches may vary from 10 to 20 degrees, but 14 to 16 degrees are the most popular.

Wind Bracing

This will be provided in the roof structure and often between the stanchions for spans greater than 9m (10yd). Often it is incorporated in the end bays of buildings only. It should also be noted that buildings that are to be used for the bulk storage of potatoes, grain and silage will also need to have extra strength to deal with the superimposed loads exerted. Buildings that are not designed and erected for this purpose will fail if these materials are stored in them (College of Estate Management, 1993).

The next figures show a typical layout of a modern farm building.

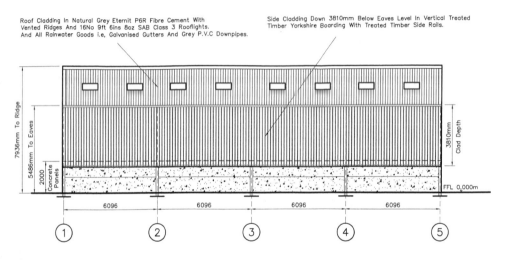

ELEVATION ON LINE B AND OPPOSITE HAND LINE A
1:100

ABOVE: A typical farm building constructed of steel, concrete, timber and fibre-cement roofing. (Shufflebottom, 2005)

BELOW: The same building as shown in the preceding figure, showing the end elevation.

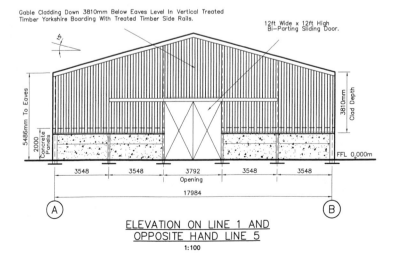

ELEVATION ON LINE 1 AND
OPPOSITE HAND LINE 5
1:100

(Note that these two figures are shown for reference purposes only and the size of foundation on the drawing is not to be used in practice; Shufflebottom Ltd will not accept any liability unless they have carried out the full calculations.)
(Shufflebottom, 2005)

A relatively modern farm building constructed of primarily steel. It is both a grain and a machinery store.

The same building as shown in the previous plate. The division between machinery and grain store can easily be seen.

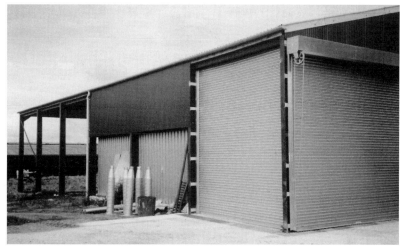

Another angle of the same building.

LEFT: Photograph of an ex-farm building (probably a grain store), now in use as an indoor tennis court.

BELOW: Outside the same fairly conventional farm building building.

BOTTOM: Another view of the same building. Notice that in the background the farmer has a small lake with fishing available as further diversification.

CHAPTER 6

Water Supply, Drainage, Electricity Supply, Lighting and Heating

WATER SUPPLY

There are many different plumbing materials found on farms, some very old and others fairly recent. Pipes may be made of lead, cast iron, asbestos cement, galvanized iron and other materials besides. Lead was very common but this has now been banned due to high lead levels occurring in water supplies. Cast iron and asbestos cement pipes are both likely to fracture through ground movement and therefore are not recommended.

Galvanized Iron

This type of pipe was common before plastics appeared and may still be specified above ground since it is rigid and cannot be chewed by livestock; this might be an important advantage in certain circumstances. On the negative side, it can be time-consuming to install, and care is needed to make sure that components can be removed for servicing without your having to remove all the pipework. All fittings will be BSP (British Standard Pipework) thread, and they are sized on the nominal inside diameter of the pipe, for example, ½in, ¾in, 1in, 1¼in, 1½in, 2in

and upwards. Threaded joints must be jointed by non-toxic joining paste or by the use of polytetrafluoroethylene (PTFE) tape; a taper thread on the pipe helps to ensure a watertight joint.

Copper

Copper may be used for certain applications, such as milking parlours, dairies and in farmhouses, both for hot and cold applications. But it can easily be damaged by livestock and can prove expensive when undertaking large installations. On the other hand, it is easy to install, does not corrode and requires little space.

Copper may easily be bent to suit the application, using a bending spring. The pipe is first cut to length by using a fine-toothed hacksaw or pipe cutter. Ends must be square and all burrs removed with a smooth file or burr brush. The bending spring is placed inside the pipe and then it is carefully pulled around the front of the knee. To prevent the spring from jamming inside the tube, pull the bend a few more degrees than required, then ease it back to its final angle; the spring may then be withdrawn easily. Where necessary, a length of strong wire can be fixed to the loop in the end of the spring so that it can be passed further along the pipe to allow bends to be made in longer lengths. Should a tight bend be required, a standard elbow fitting must be used.

Common pipe diameters (always measured as the outside diameter) are 10, 15, 22, 28 and 35mm. The pipe ends may be joined either with compression fittings or soldered joints. 'Yorkshire' fittings contain their own solder, put in by the factory, and therefore provide an easy to make joint. The following figures show methods used to connect pipes.

Clean the ends of the pipe and the inside of the fitting with wire wool or burr brush until the surfaces are clean and bright.

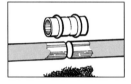

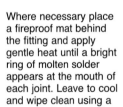

Spread a thin film of flux over the pipe ends, and push pipe into the fitting until it stops against the internal shoulder. All joints on each fitting should be completed at the same time.

Where necessary place a fireproof mat behind the fitting and apply gentle heat until a bright ring of molten solder appears at the mouth of each joint. Leave to cool and wipe clean using a

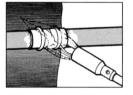

dry cloth. Soldered joints can be separated by heating and drawing pipe out while solder is molten.

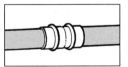

The completed joint.

Capillary fitting. A solder joint made by heat, which melts a solder ring inside the body of the fitting fusing the fitting to the pipe. (Payless DIY, 1986)

Proceed in the same way as with capillary fittings. As the joint is heated apply solder. When solder has stopped being drawn up remove heat and allow to cool before wiping clean. With the end feed individual joints can be soldered providing they are well fluxed before heating.

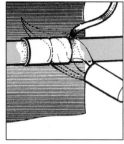

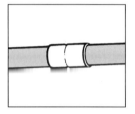

The finished joint.

ABOVE: *End feed fitting. A solder joint made in a similar way to a capillary joint. The difference is that solder is not contained within the body of the fitting, it is applied separately as the fitting is heated. (Payless DIY, 1986)*

Tools required for capillary and end feed fittings: gas torch, fireproof mat, steel wool, flux and, for end feed fittings, a spool of solder. (Payless DIY, 1986)

Assemble nuts and olives onto pipe as illustrated.

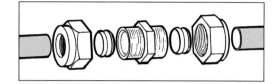

With the pipe ends in firm contact with the shoulder inside the fitting, hand tighten both nuts and then spanner tighten a further ½ turn for 15 mm fittings and ¾ turn for 22 mm fittings. If required fittings can be separated by unscrewing the compression nuts.

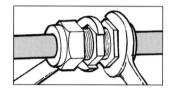

The completed joint.

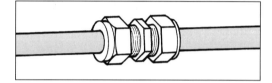

Compression fitting. A mechanical joint quickly fitted and requiring only two spanners to complete the joint. (Payless DIY, 1986)

Tools required: two spanners or adjustable pattern.

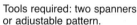

Polythene

This is a plumbing material commonly used on farms. Pipes that are underground are invariably of polythene; pipes should be black for above ground and blue for below ground, to distinguish them from electricity and gas pipes. All pipes and fittings are now metric. Blue piping is ultra-violet-light degradable and should therefore be shielded from any direct sunlight if it is used above ground for any purpose, for instance, stretching to the ball valve of a water drinking trough. Medium density polyethylene (MDPE) is occasionally used and will withstand a pressure of 10bar. Fittings that used underground on polythene pipe must be of gunmetal or another type resistant to corrosion (dezincification). Resistant fittings are often marked DR; sometimes DR fittings are a

necessity above ground also because of the nature of the water supplied.

Water companies will often use plastic fittings on smaller supplies; the Polygrip fittings (invented by a farmer) are used extensively. They can be fitted to a range of obsolete pipes and fittings, including lead pipes. Fittings must be tightened with a spanner, as normal, but not over tight since this can lead to the cracking of the body of the fitting. There are other types of fitting used for underground pipes that are push-fit only; they do not require spanners for tightening. Plastic pipes may also be made of unplasticized polyvinylchloride (PVC), but this is used only on waste pipes. The fittings may be either glued or simply pushed together. Since the pressure that they will have to withstand is near zero this method is perfectly acceptable.

DRAINAGE SYSTEMS

A drainage system on a farm will be for one of the following purposes:

- a *domestic foul* system that collects sewage waste from toilets and conveys this either to a public sewer (unlikely in agriculture) or to a septic tank
- a *dirty water* system that has lightly polluted water, for example, from dairy parlour washings, that may be transferred to settlement tanks before a low-rate irrigation system for example
- a *slurry system* which is used for animal waste; this may be treated, stored and eventually spread on to fields
- a *clean water* system which collects roof water and other non-polluted sources and conveys this to a stream.

Septic Tanks

These systems are able to accept material which is then digested by bacteria over time in anaerobic conditions, until it is safe to discharge. They are often moulded as a one-piece, corrosion-resistant, polyethylene tank. This type of construction ensures a tough, impact-resistant, leak-proof tank that is usually fitted with sturdy lifting eyes to make lifting and installation easier. Tank sizes may vary considerably, but one company offers 2,800 to 6,000ltr (616–1,320gal) capacities. Tanks may be installed in either a 'dry' ground condition or a 'wet' ground state (Hepworth Drainage, 2000).

For a single house installation, 15m (49ft) away is generally regarded as the minimum distance, although approval and agreement must be sought from the controlling body at an early stage. The direction of the prevailing wind should also be taken into account when considering the ideal siting; the tank should not be situated close to a driveway, roadway or anywhere where there is a risk of its being subjected to additional superimposed loads. Good access should be provided for the sludge emptying tanker. Often an effluent soakaway system may be utilized with a septic tank, so that the need for 'emptying' becomes automatic (Hepworth Drainage, 2000). The next set of figures show diagrams of typical underground drainage systems.

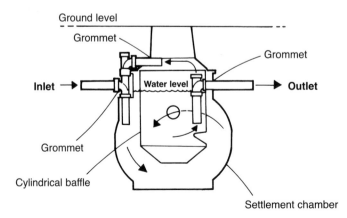

Sketch showing the layout of a polyethylene septic tank. (Hepworth Drainage, 2000)

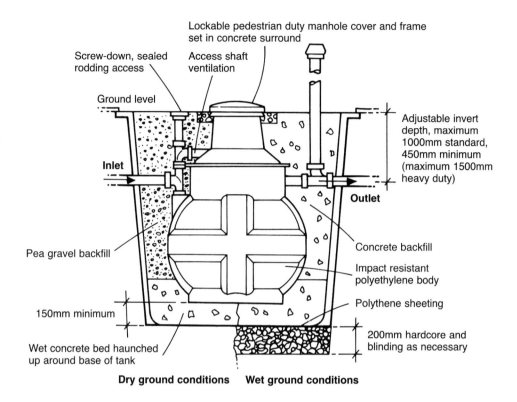

Lockable pedestrian duty manhole cover and frame set in concrete surround

Screw-down, sealed rodding access

Access shaft ventilation

Ground level

Inlet

Pea gravel backfill

150mm minimum

Wet concrete bed haunched up around base of tank

Adjustable invert depth, maximum 1000mm standard, 450mm minimum (maximum 1500mm heavy duty)

Outlet

Concrete backfill

Impact resistant polyethylene body

Polythene sheeting

200mm hardcore and blinding as necessary

Dry ground conditions   Wet ground conditions

ABOVE: Detailed diagram depicting the layout in (left) dry ground and (right) wet ground conditions for a septic tank. (Hepworth Drainage, 2000)

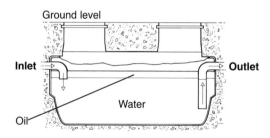

Ground level

Inlet   **Outlet**

Water

Oil

ABOVE: An oil interceptor (designed to meet BS 8301) for 'high risk' areas where there is heavy oil contamination. (Hepworth Drainage, 2000)

RIGHT: Diagrams showing how an effluent distribution soakaway system should function. (Hepworth Drainage, 2000)

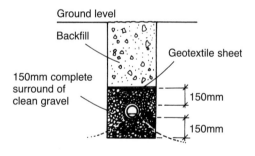

Ground level

Backfill

Geotextile sheet

150mm complete surround of clean gravel

150mm

150mm

Soakaway kit supplied with HST 27 septic tank

Septic tank   Air vent

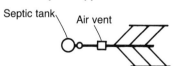

Soakaway kit supplied with HST 38 septic tank

Septic tank   Air vent

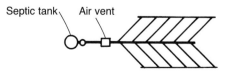

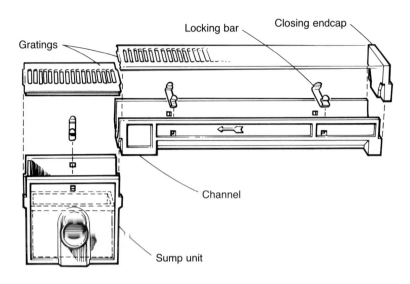

A rain drain system designed to clear away surface water. Note that this system is for surface water and not effluent. (Hepworth Drainage, 2000)

Pipes

Pipes that are used underground may be made of cast iron, glazed stoneware, extruded clay, concrete or plastic (uPVC). Many of these materials are no longer used to a large extent, but plastic has become very popular. The possible applications are numerous and may include gravity sewerage systems, surface water drainage, catchments and chambers, land drainage, ducting systems and others such as crop ventilation, animal drinkers and rainwater harvesting. Many of the larger plastic pipes, used for drainage systems, have advantages such as:

- lighter than concrete pipe (approximately 6 per cent of the equivalent concrete pipe)
- good resistance to high water pressure longer lengths available with integral sockets for connections
- jointing systems that will remain intact, even under extreme site conditions
- robust, flexible construction with long life
- good chemical, impact and abrasion resistance
- immunity from sulphate attack or corrosion due to sewer gases
- coloured internal walls which make for ease of CCTV operation.

Plastic pipes may be jointed either by glue, which gives a permanent fixing (this is used on small bore drainage systems for example) or simply by push fit. The pipe end is pushed into a joint that has an O-ring that seals the joint. The advantages of this method are that the pipe can be dismantled again for repositioning and reassembling. The other type of fitting that is common is the dry seal gasket, as used on most guttering systems. Here, the different sections of rainwater goods are simply clipped together in order to make a watertight seal.

Sumps

Sumps or underground tanks can be useful for collecting effluent, slurry and generally preventing problems of pollution. Very large tanks easily become problematic to construct, and certainly the larger they are the more expensive they become. Often they will be constructed using a 150mm (6in) reinforced concrete base and then built up with bricks, blocks (reinforced and rendered), reinforced concrete or precast concrete rings (or panels). If the tank is below the size 4.5m × 4.5m × 2.5m deep (18ft × 18ft × 8ft) it can be constructed by farm staff; on the other hand, if it is bigger than this it will require an engineer to supervise. An important consideration is ground water pressure, which will lead to the flotation or collapse of walls when the sump is empty.

Surface Drainage

Probably the most effective way of taking surface water away from a yard is to have correct surface drains fitted. One such system uses a sump unit, with connector to sub-surface or soakaway pipe sewer, and lengths of concrete channel 1m (3ft) long with a grating together with locking bar. These may be used around buildings or for carrying the water away from roads and aprons (see the figure opposite).

Oil Interceptor

Interceptors are used to stop oil and other hydrocarbons from entering a drainage system and therefore causing pollution. They are usually made from medium-density polyethylene to give a construction that has impact resistance, durability and is a one-piece unit, thereby offering a leak-proof system. One such system is available in sizes from 1000–8000lt (220–1760gal). Units may be full-retention or by-pass interceptors, or the type of interceptor that is an automatic closure device which shuts off the drain when the interceptor becomes full. (see the diagram of an oil interceptor on p.75 (middle)).

Reed Beds

A reed bed is a natural filtration process which may be used in conjunction with a Bio-tech (Titan, 2004) system in order to treat sewage. This system may be used with camping and caravan sites, country clubs and restaurants, for instance, where there is no connection possible to a mains sewer. A reed bed would be required only when a local water authority requests a better quality of effluent than that discharged from a standard unit. The entire system would comprise: primary settlement stage, carbonaceous biological treatment, nitrifying biological treatment and final settlement – possibly to a reed bed. It is a method that no doubt become more popular in the future.

Gutters and Downpipe Sizing

A method may be used to assist in the design of eaves gutters and downpipes; this is shown below. For this calculation, all gutters are assumed to be laid level.

1. Calculate the effective roof area to be drained per m run of eaves; A = width of the building, mono-pitched roof or half the width if a ridged roof

$$A \underline{\quad} m$$
$$= (a \underline{\quad} m \times \tan (\text{roof pitch}) \underline{\quad} \times 0.5)$$
$$= \text{effective area} \underline{\quad} \text{sq m}$$

2. Calculate the rainwater run-off; the design rate for rainfall is usually 75mm/hr.

> effective area ____sq m
> × rainfall ____mm/hr ÷ 3600
> = run-off ____ltr/sec

3. Choose a gutter type and size and read off flow capacity from the table below; an arbitrary choice will have to be made in the first instance

> gutter size ____mm gutter type
> flow capacity ____ltr/sec

4. Calculate the maximum gutter length from an outlet

> flow capacity length ____ltr/sec
> ÷ run-off ____ltr/sec
> = gutter length ____m

If the gutter length is greater than the total length of the building × 1.1, then go back and choose a gutter with a smaller flow capacity, the gutter will overflow if any part of it is greater than the gutter length from an outlet; if the answer is less than 1 then only one outlet is required.

5. Calculate the maximum outlet spacing

> gutter length ____m × 2
> = outlet spacing ____m

6. Calculate the number of outlets required

> roof length ____m ÷ outlet spacing ____m
> = number of outlets ____

7. Choose an outlet diameter from the table opposite

> outlet type ____
> downpipe diameter ____mm

Your choice of gutter and downpipe are now found within the calculations above (FRBC [n.d.])

Gutter flow capacities for several types of material

Nominal gutter size mm	True half-round gutter (1, 2, 5)	Nominal half-round segment (3, 4, 5)	Ogee gutter (2)	Ogee gutter (3, 4)	Semi-elliptical (5)
75	0.4	0.3	–	–	1.83
100	0.8	0.7	0.9	0.5	–
115	1.1	0.8	1.4	0.7	–
150	2.3	1.8	2.6	–	–

1. Asbestos cement to BS 569
2. Pressed steel to BS 1091
3. Aluminium to BS 2997
4. Cast iron to BS 460
5. Unplasticised PVC to BS 4576

(FRBC (n.d.))

Downpipe diameters for several dimensions of gutter size			
True half-round gutter size	Sharp or round-cornered outlet	Outlet at end of gutter	Outlet not at end of gutter
75	sc	50	50
	rc	50	50
100	sc	63	63
	rc	50	50
115	sc	63	75
	rc	50	63
125	sc	75	89
	rc	63	75
150	sc	89	100
	rc	75	100
			(FRBC (n.d.))

ELECTRICITY SUPPLY

Below are some of the terms commonly used.

Current – the flow of electricity in a wire, measured in amperes (A); the symbol for current is I.

Voltage – the pressure which moves electricity through a wire, measured in volts (V); mains electricity is normally at a fixed voltage of 240V.

Resistance – the resistance to the flow of electricity, measured in ohms (Ω); materials differ in their resistance to the flow of electricity: poor resistors are good conductors (gold, copper, steel, carbon, water solutions) and good resistors are poor conductors of electricity or insulators (rubber, plastic, glass, porcelain, mica, air); resistance is shown as heat which results in power wastage; the unit for resistance is the ohm and Ohm's law gives the formula $R = V/I$.

Electrical power – measured in watts (W); the symbol for power is P; 1000W = 1kW; electrical power is found by multiplying volts and amps, that is, $P = V \times I$; to calculate a fuse size, we can say that $I = P/V$.

Types of Electrical Current

Direct current (dc) flows continuously in one direction only. A dynamo on an old tractor produces dc, and dc is used on tractors, cars and small generating plants. A battery gives out dc and requires dc for its re-charging.

Alternating current (ac) flows first in one direction and then in the other. This change in direction is very rapid and in Britain it occurs at 50 times per second, or 50Hz. Another name for this is the frequency. Power stations produce ac and this current is used in the mains supply to industrial, farm and domestic properties.

The voltage (pressure) of ac can readily

be 'boosted' for transmission over long distances. Thus ac is always used where electricity is produced on a large scale for distribution over large areas – the national grid will operate at 132000V, whereas local grids normally operate at 33000 or 11000V.

Mains Electricity Supply

Current may be supplied to the consumer in two ways:

Single phase – the voltage supplied is 200–250V. A single alternating current passes through the appliance and the voltage rises to + 240V then decreases to nothing; the voltage drops still further to – 240V, that is, the 'pressure' is in the opposite direction and it finally rises to zero. This completes the cycle which occurs 50 times per second. The supply is adequate for small kilowatt motors and for normal heating and lighting purposes.

Three phase – the voltage supplied here is 400–440V. Three separate, single-phase currents which are out of step (maximal and minimal voltages occur at different times) are fed to the appliance. This results in a total voltage of approximately 415V. Since the phasing is staggered, three-phase electricity can offer advantages for heavier loads, for example, large electric motors are normally of the three-phase type.

Electricity that is to be transmitted over a long distance is usually converted to a very low current, and, incidentally, to a very high voltage. Since the equation for power loss is (current)^2 × resistance, the lower the current, the less the power loss. The conversion of voltage and amperage is carried out by a transformer.

Types of Supply Cable System

The mains current from the supply may be carried by a two-, three- or four-cable system, for example: the two-cable system for a single phase supply, the three-cable system is used for a three phase supply and the four-cable system (which has three cables carrying current, the fourth being a neutral or return wire) is used for three-phase and single-phase, supplies. Note that the colours of the live wires in a three-phase wire are red, yellow and blue while the neutral is black, but in March 2006 the colours of mains wiring were changed.

Fuses

A fuse is a special metal strip of limited current-carrying capacity (similar to a shear bolt in a power-driven machine) and it is used as a safety device to protect a circuit from electrical overloading. That is, if the current flowing in a circuit rises above normal owing to a fault having developed, for instance, a short circuit, the fuse will melt (or blow) and break the circuit. The following are examples of common faults: the circuit has been overloaded, for example, by the use of too many appliances at one time or the use of an appliance which consumes too great a current; an appliance is faulty, for example, 'shorting' between two wires; or there is dampness in the circuit or appliance causing a 'short'.

Fuse wire and fuse cartridges are made in several sizes, 2, 3, 5, 10, 15 and 30A being fairly common ratings. This will mean that a 2A fuse, for example, will safely carry 2A before it will melt; a lighting circuit will have a 5A fuse, for example; an immersion heater will have a 15 or 20A fuse; a ring main (socket outlets) and average size cooker 30A; a large cooker might be 45A. It is extremely important that a fuse is replaced with one of the correct value in order for the circuit to carry the right amount of current. For safety, it is always essential to switch off at the mains before replacing a blown fuse. Typical questions that might arise are: what size of fuse should be fitted to the following circuits? A heater of 2kW running from the mains (240V) supply (8.3A, therefore fit a 10 or 13A fuse), or a television of 600W running from a 250V supply (2.5A, therefore fit a 3A fuse).

Wiring a Plug

The first thing that should be done is to inspect the plug for damage – cracked or burned insulation, bent or corroded pins, damaged screws or fuse holder, damaged or missing cord grip. The flex must be checked to see that it is in a serviceable condition, that is, the outer insulation is not cut through, and the 'other end' correctly attached and functional. It is always wise that you should not fit a plug unless you know that the appliance is safe to use and of the correct voltage for the plug/supply. And the flex must be of the correct size and type for the particular application and with the correct insulation. Note that it is illegal to sell equipment with an incorrectly coloured flex.

Essential points to mention are set out below.

1. flex trimmed to the correct lengths; the earth wire should normally be made longer than the other two
2. inner insulation undamaged, with no wire showing, including the earth wire
3. no wire strands cut or damaged
4. all wires twisted together

5. all wires firmly attached
   - clockwise around terminal or clamped under screw (thin wires folded double); no crossed threads
   - no trapped insulation
   - no stray strands
   - no excessive lengths of bare wire
6. all wires attached to the correct terminal
7. outer insulation correctly clamped in cord grip
8. fuse correctly located (firm in the spring clips)
9. fuse of the correct size (amperage) and type
10. plug top correctly fitted (located in the correct position and fully home, with no trapped flex).

Note: older plug tops were located on the flex first and were not 'cut' at the point of flex entry and newer appliances tend to be fitted with plugs as an 'all-in' design, thereby eliminating any DIY attachment.

Reading a Meter

There are two main types of meter which may be present on a holding: digital meter and dial meter. Digital meters are the modern type and the number of units (kWhr) used is shown by a row of figures. Always remember that the reading on a digital meter is the total number of units recorded. If and when this number exceeds 99999 the meter starts again at 0 (like the mileage indicator on a car). Sometimes there will be two rows of figures; the top one will show the off-peak electricity and will be marked 'low'. This is where the farm will be charged at a lower rate for using electricity at night, for example, for the milk cooling system. The bottom row shows the normal rate units and is marked 'normal'.

Several types of lamp that may be found in agricultural buildings. Left to right: tungsten filament, tubular fluorescent, high-pressure sodium and metal halide.

A dial meter should be read right to left. Always remember that two adjacent dials revolve in opposite directions. Ignore the first reading, 1/10, since this is present only for testing purposes. Start by reading the single units, then tens of units, then hundreds, then thousands and finally tens of thousands. Always write down the number the pointer has passed, so if, for example the pointer was between 3 and 4, and then write down 3; if it is directly over a figure, say 7, look at the pointer immediately to the right – if this is between 9 and 0, write down 6, if it is between 0 and 1 then write down 7.

LIGHTING SYSTEMS

There are many different lamps that may be applicable in agriculture. In pig and poultry houses, for example, probably the most common are the tungsten and the fluorescent lamp; these will be described along with other discharge lamps that may be suitable for farming.

Note that the lumen (lm) is the unit of luminous flux, which is the rate of emission of light, derived from the radiant flux by weighting the radiation with respect to the sensitivity curve of the human eye (Philips, 1984), and 1lm/sq m is equal to one lux (lx). A typical 240V, 100W bulb has a lumen output of around

1,250lx, compared with bright sunlight at 80,000lx.

Lamps are rated by their efficacy or by how much light (in lumens) is produced by 1W of electrical power. Typical values are:

Lamp type	Efficacy (lm/W)
tungsten filament	8–15
tungsten halogen	16–22
compact fluorescent	40–50
standard fluorescent	52–70
HF fluorescent	89
sodium discharge (high pressure)	54–107
sodium discharge (low pressure)	70–170

(Farmers Weekly, 30 Dec. 1994)

Tungsten Filament (GLS)

A tungsten filament lamp is heated to incandescence inside a glass envelope, producing white light. Most of the energy input is emitted in the form of heat (which is why they get very hot), and hence the lamps operate at low efficacy. The operating life of these lamps is about 1,000hr. As far as their energy efficiency is concerned, this is poor with typically 8–18lm/W. They are very common and are used as a general purpose light. Often covers will be required for these lamp fittings in order to keep broken glass away from animals. Usually the lamps will have dimmers fitted.

The advantages of this lamp are: instant switching on and off, excellent colour rendering, versatile, cheap to buy and can be dimmed. The main disadvantages are: poor efficacy, high running costs, short operating life and high surface temperature.

Fluorescent Lamp

This lamp contains mercury vapour at low pressure. An arc emits ultra-violet energy which is converted into light by the coating of fluorescent phosphors on the tube (BRE, 1993). High frequency fittings are available at higher cost, which improve energy efficiency and provide a dimming facility; these are usually smaller diameter tubes. Standard fluorescent tubes (MCF/U T12) are 38mm (1.5in) in diameter, whereas the low energy (MCF/U T8) varieties are 26mm (1in) in diameter. The energy efficiency is relatively high, with 37–100lm/W. They may not be suitable for laying hens because of their 'disco effect'. This type of lighting induces stress, feather pecking and cannibalism (which do not occur with incandescent lights).

The advantages of this lamp are: high efficacy, low cost, cheap to run, a good range of tube colours; good to excellent colour rendering (depending on the specification of fluorescent coating), life is typically 5000–15000hr and a diffused light which reduces the amount of shadows. The disadvantages are: lower output per lamp compared to some other types (such as the high-pressure sodium type), high-frequency cold start gear may be required in low temperatures, not as suitable for

high bay installations and cannot be dimmed without special control gear. The lamp has a very wide range of uses, including inside piggeries.

Compact Fluorescent

This is a relatively new type of lamp and one that will normally go straight into the fitting of a tungsten filament bulb. The wattage may be very low, for instance, a 12W compared with a 60W conventional bulb. The big advantage with this type of lamp is its low running cost with a very long life, but the disadvantage is the higher initial cost; it is not suitable for dimming, and is somewhat bulkier.

Discharge-type Lamps

There are several other discharge-type lamps that may be suitable for certain applications. The high-pressure sodium lamp operates with an arc formed in high-pressure sodium vapour. It has a slight delay in warming up, say, 5–7min ignition time with a restrike time of 1min. It has a very high efficacy, typically 67–1,37lm/W (BRE, 1993). The lamp has a long life, 6,000–24,000hr, with very little fall off in light output with age. One of the main drawbacks is its higher cost.

The metal halide (MBIF) type is an electrical discharge in a high-pressure mercury atmosphere with metal halide additives in an arc tube, sometimes contained within a glass envelope. Ignition time is 5min, with a restrike time of 10min. It has a high efficacy, typically 66–84lm/W and its life is usually 4,000–12,000hr. It is mainly used for indoor and outdoor floodlighting.

The tables overleaf show lamp types and lamp suitability.

Lamp types that may be found in agricultural applications and their main advantages

Lamp type	Tungsten Filament	Tungsten Halogen	Compact Fluorescent	Fluorescent Tube	High Pressure Sodium	Metal Halide	Mercury Vapour	Low Pressure Sodium
	GLS	TH	MCF	MCFU	SON	MBIF	MBF	SOX
Warm Up Time	Nil	Nil	1–2 secs	1–2 secs	3.5–5 mins	3.5–5 mins	3.5–5 mins	8–10 mins
Efficacy lm/w*	10–18	16–22	35–70	45–72	52–105	62–102	30–50	90–142
Rated average life	1,000	2,000	10,000	9,000/15,000	28,500	14,000	20,000	16,000
Colour	White (Other colours available)	White	White	White (Other colours available)	Warm white	White	Blue/White	Yellow
Colour Rendering	Fair/Good	Good	Good	Fair	Good	Fair	Poor	Good
Power Range (W)	15–1000	100–2000	5–55	15–125	50–1000	100–2000	50–2000	18–180
Starting Gear	No	No	Yes	Yes	Yes	Yes	Yes	Yes
Other comments	Lamp life shortened by vibration, overheating and high voltage	Use fittings with a glass front		Lamp life shortened by frequent switching. Output reduced significantly by low temperatures				

(Farm Electric, 1996)

* Note – for any particular lamp type, larger lamps (wattage) are more efficient

Lamp suitability for ten agricultural applications, with some recommendations

Location	Tungsten Filament (GLS)	Tungsten Halogen (TH)	Compact Fluorescent (MCF)	Fluorescent Tube (MCFU)	High Pressure Sodium (SON)	Metal Halide (MBIF)	Mercury Vapour (MBF)	Low Pressure Sodium (SOX)	Recommended option
Milking Parlours	X		X	XXX					Fluorescent tubes with splash- or waterproof fittings
Cubicle Sheds	X	XX	XX	XXX	XXX	XXX	XXX	X	Splash- or waterproof low bay HP Sodiums for roof mounting for low running costs. **Fluorescent lights** with splash- or waterproof fittings for good distribution of light at lower illuminances.
Piggery	X		XX	XXX	X				**Compact fluorescents** are a good replacement for existing tungsten lights but cannot be dimmed. **Fluorescent tubes** can be dimmed using special control equipment. Use sealed waterproof fittings. For high yards use **HP sodiums** as with cubicle houses.
Poultry	X		XX	XXX					**Compact fluorescents** are a good replacement for existing tungsten lights but cannot be dimmed. **Fluorescent tubes** can be used with special dimmers. Use sealed waterproof fittings.
Grain Stores		XX		X	XXX	XXX	XXX		**Low bay HP sodiums** For instant light use tungsten halogen, but beware that wall mounted fittings can cause dazzle.
Potato Stores		X		X	XXX	XXX	XXX		For Bulk stores see 'Grain Stores' information. For boxed stores, use **high bay HP sodium** fittings over 'pathways' to provide localised lighting.
Packhouses				XXX	XX	XXX	XX		Tubular fluorescents for good even light distribution and low bay metal halide for higher buildings. Both give good colour rendering. For less critical colour needs HP sodium is suitable. Use 'North light' type tubes for excellent colour rendering over sorting tables.
Workshops			XXX	XXX	XX	XX	XX		Fluorescent tubes give least shadows and even light. Use high frequency lights to avoid 'strobing' on rotating machines. Tungsten bulbs for close work or small tubular fluorescents
Yards		XX			XXX	XX	XX	XX	HP sodium floods for long term lighting. Tungsten halogen floods with 'sensors' for short term lighting. Position carefully to stop dazzle.
Security		XX	X		XXX			XXX	Tungsten halogen floods with 'sensors'. Low pressure sodiums are the most economic light.

Key: the more X's the better

(Farm Electric (1996))

HEATING SYSTEMS

Again, there are many different heating systems that may be used for various applications in agriculture. In pig and poultry houses, probably the most common are space heaters, run from gas or oil, electric infra-red heaters and under-floor (electric) heating systems. These will now be described more fully.

Warm Air Heaters (Floor Standing or High-level Units)

These units are generally the cheapest to buy and install. Unflued units burn natural gas, propane or oil directly in the air supplied to the space. Flued units burn the fuel in a heat exchanger, warming the supply air which is then discharged horizontally via a fan. The air discharge louvres are adjustable, allowing the air flow to be directed as required. Alternatively, the supply air can be distributed via ductwork. The combustion products from the heat exchanger are discharged to the outside via a flue or chimney. Occasionally, the use of fans is recommended to assist air circulation and resist air stratification occurring. A combustion air inlet in the building fabric must be provided for flued units.

Flued units operate at approximately 75–80 per cent efficiency (85–90 per cent efficiency for unflued units). Modulating

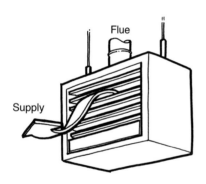

High-level unit

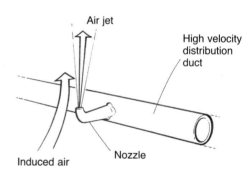

Warm air jet/induced jet system

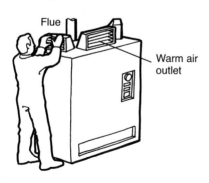

Floor-standing warm air heater

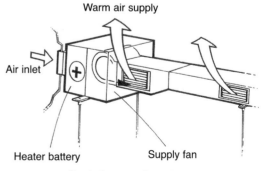

Ducted warm air system

Some types of convective warm air heater that may be found in agriculture.

burners are available which match heat output to heat load, hence reducing fuel consumption.

The main advantages are: low capital cost, low installation cost, flexibility (units can be moved relatively easily to suit new house layouts) and good access for maintenance. The main disadvantages are: floor space required (unless a high-level unit is used, often suspended from a chain/cable), predominantly convective heat, hence stratification can occur especially in high buildings, high dry bulb temperatures required to achieve same comfort level compared with radiant systems but increases energy consumption, air movement required to distribute heat (this can cause uncomfortable zones adjacent to the discharge of high velocity hot air), noisy and vulnerability to damage.

A view of a warm air heater that could be used for pig or poultry housing. (Maywick, 2004)

Gas Brooders

Brooder heaters are designed for broilers, pullets and duck-brooding layouts. The diagram shows the basic layout of one unit; it has a gas burner, a reflector to direct the heat, a gas supply and the relevant control gear. This might vary from a manual high/low control by using a simple tap, to automatic high/low incorporating a non-electric thermostat which adjusts the gas flow between fully on and partly on. A control knob provides the temperature setting. Another version is an automatic on/off model. This incorporates a non-electric snap-acting thermostat and a permanent pilot. A control knob provides the temperature setting causing the brooder to be fully on or extinguished. The figure below depicts a gas brooder that may be suitable.

The 'bird pattern' diagram overleaf

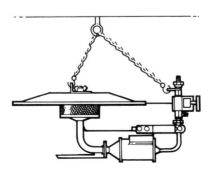

A gas brooder that would be applicable for heating in a poultry house.

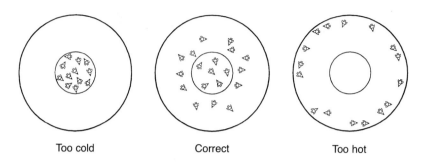

Too cold	Correct	Too hot

Bird distribution under different heating levels. A diagram to show how poultry will congregate (or otherwise) when the house temperature varies.

shows how the settled birds demonstrate if the brooder (or any other type of heating system) is working as it ought. If the building is too cold, the birds will huddle together; if it is too hot the birds will disperse to the outer walls of the enclosure. A correct heat will show that the birds are evenly distributed around the building. The figure above shows how the bird pattern demonstrates heating efficiency.

Electric Radiant Heaters

These comprise quartz lamps fitted in a polished reflector. The lamps emit radiant heat which can be directed to the area required. Heat up to full power is instant in switching on. Mounting heights can vary from 2m (6.5ft) (or less) to 4m (13ft). The lamps operate at 100 per cent efficiency of delivered electrical energy input, but use standard rate electricity. A 'black bulb' thermostat senses the radiant temperature and provides control to the heater, switching lamps on and off as necessary. These heaters have been used in many farming situations, including workshops, dairy parlours and calf rearing houses.

The advantages of this heating are: fast heat-up rate, small and lightweight,

simple and cheap to install, high level mounting – no floor space is used and thus the occurrence of accidental damage is minimized, low maintenance, no air movement, easily relocated to suit new building layout and no fumes produced. The disadvantages are: standard rate electricity is used and high running cost can result with high heat load; it is for this reason these heaters have not become as popular in agriculture as perhaps they otherwise might have done.

Underfloor Electric Heating

Electric underfloor heating – creep heating – will provide an even floor temperature and a warm, clean and stable environment which can be accurately controlled. Systems may also have timers so that further controllability may be incorporated. This system therefore offers improved animal growth and feed utilization. The diagram on p.90 shows the layout of a typical system; the cables are laid on to the insulated concrete floor using a fixing strip, which is then covered with a sand and cement screed. In situations where there is an existing floor a special thin insulating board can be laid, the cables are then attached direct to the

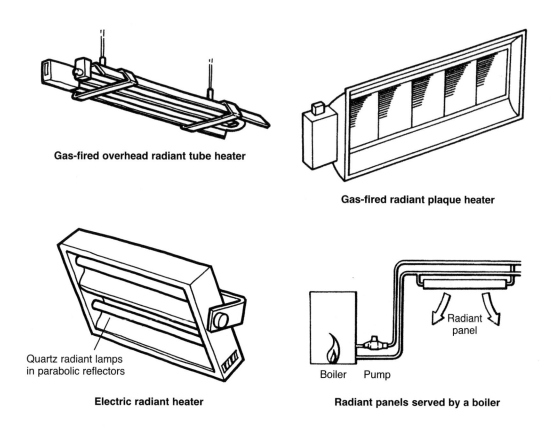

Gas-fired overhead radiant tube heater

Gas-fired radiant plaque heater

Quartz radiant lamps
in parabolic reflectors

Radiant
panel

Boiler   Pump

Electric radiant heater

Radiant panels served by a boiler

Radiant heating systems that may be used in agricultural applications.

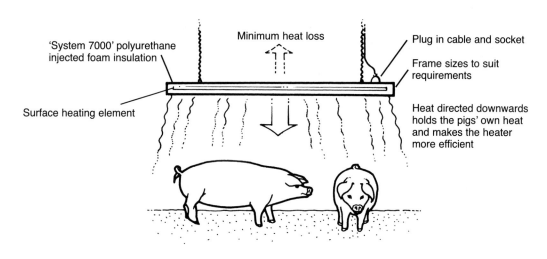

'System 7000' polyurethane
injected foam insulation

Minimum heat loss

Plug in cable and socket

Frame sizes to suit
requirements

Surface heating element

Heat directed downwards
holds the pigs' own heat
and makes the heater
more efficient

An overhead electrical warm air heating system used in a piggery. (Pyramid, 1996)

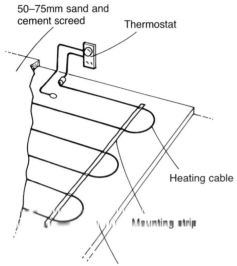

50–75mm sand and cement screed

Thermostat

Heating cable

Mounting strip

New build: 150mm 'Lytag' insulating concrete
Renovation: 'Wedi' insulating board

An underfloor heating system that may be applicable in a piggery. (DEVI, 2006)

Provided that the floor is laid and terminated correctly, the heating system has a lifespan as long as the building's and is also durable, which means that the floors can be cleaned with a high-pressure hose without having to remove any electrical equipment. The figure opposite shows a typical underfloor heating system.

An alternative system is to use heating pads. These may be in various sizes, for instance, 1.2m × 0.4m (4ft × 1.3ft), 1.2m × 0.6m (4ft × 2ft), and so on. Their life expectancy is almost indefinite in normal working conditions and they have a high mechanical strength. Urine and water do not present a problem to the pads and they are triple-insulated for safety. The pads can be embedded in concrete to give a level floor. In trials, pads have held the temperature remarkably steady, even with outside temperatures as low as -2°C, giving a better food conversion ratio as well as low running costs (Riverina, 1995). The main advantage of this type of heater as opposed to the individual cables is that it can be laid as one unit without all the trouble associated with cables. There should be far less room for human error with this type of system.

board which is then covered with the screed. Both applications use heating cables which are double-insulated and typically available in 15m (49ft) lengths with loadings ranging from 155 to 4260W, thus allowing for individual or multiple creep heating control.

CHAPTER 7

Heat

HEAT ENERGY

Heat, usually given the symbol H or Q, is a form of energy. The standard SI unit used for heat is the joule (J). The other main unit that is used is the kilowatt-hour, where 1 kWhr = 3.6 MJ. There are other forms of energy apart from heat, but all can be converted to thermal energy. For example, combustion (burning) converts the chemical energy contained in the materials to heat. Interestingly, the thermal energy radiated from the sun is also the origin of practically all energy used on earth, for example, fossil fuels such as coal and oil, which were originally forests grown in the presence of sunlight.

TEMPERATURE

Temperature is often confused with heat but they are not the same thing. 'A red-hot spark, for example, is at a much higher temperature than a pot of boiling water; yet the water has a much higher "heat" content than the spark and is more damaging' (McMullan, 2002). 'Temperature is the condition of a body

that determines whether heat shall flow from it' (McMullan, 2002). The unit of temperature is the kelvin (K), formerly the degree Kelvin.* Heat will flow from objects which are at a high temperature to objects at a lower one; when there is no flow of heat between two objects the conclusion is that they are at the same temperature. A thermometer is used to measure temperature by using some property of a material that changes in a regular manner with a change in temperature. Some common types of thermometer are: mercury-in-glass, alcohol-in-glass, thermoelectric thermometers, resistance thermometers and optical thermometers.

HEAT CAPACITY

It is true that the same mass of different materials can 'hold' very different quantities of heat. For example, water has to be supplied with more heat compared with oil to produce the same rise in temperature. Therefore water has a greater heat capacity than oil. The specific heat capacity of a material is the quantity of heat energy required to raise the temperature

*In the Kelvin scale measurements are made from absolute zero, thus the freezing point of water is 273.15K; the temperature intervals on the Kelvin and the Celsius scale are identical.

of 1kg of that material by 1 Kelvin or 1°C (McMullan, 2002). The unit for the specific heat capacity is J/kgK. Specific heat capacities for some common materials are given in below.

Material	Specific Heat Capacity (J/kgK)
water	4190
concrete and brickwork	3300
ice	2100
paraffin oil	2100
wood	1700
aluminium	910
marble	880
glass	700
steel	450
copper	390

Note: the values for particular building materials vary (McMullan, 2002)

Since the heat capacity of water is higher than the values for the other materials, water is a good medium for storing heat; heat exchange devices such as boilers and heating pipes utilize the high heat capacity of water to transfer heat from one place to another.

CHANGE OF STATE

Within the normal range of temperature and pressure, there are three states of matter; these can be illustrated as:

- solid state: the molecules are held together in fixed positions; the volume and shape are fixed
- liquid state: the molecules are held together but have freedom of movement; the volume is fixed but the shape is not

- gaseous state: the molecules move rapidly and have complete freedom; the volume and shape are not fixed. (McMullan, 2002)

The state of a substance will depend upon the temperature and pressure present at the time, for instance, if water is considered, then it will be appreciated that at certain temperatures it undergoes a change in state and its energy content is also increased or diminished. The following shows the release of heat from a gas or a liquid:

gas (condensation)
→ liquid (solidification or fusion)
→ solid

$$H = mc\theta$$

where:

it is the quantity of sensible heat (J)
m is the mass of substance (kg)
c is the specific heat capacity of that substance (J/kgK)
θ is the temperature change (°C)

SENSIBLE HEAT

This term refers to the rise in temperature of a substance as heat is supplied to it; the word 'sensible' is simply used because it is apparent to our senses. Sensible heat is the heat energy absorbed or released from a substance during a change in temperature (McMullan, 2002).

LATENT HEAT

When a substance is changing from one

state to another, such as from a solid to a liquid, the temperature remains constant even though heat is being supplied. This heat is called 'latent' because it appears to be hidden. Latent heat is the heat energy absorbed or released from a substance during a change of state with no change in its temperature (McMullan, 2002). The latent heat is considerably greater than the amount of heat required to raise water temperature from cold to boiling point. The latent heat of steam is, in fact, 70–80 per cent of the total heat. When steam gives up its heat it loses its latent heat until it condenses, the condensate being at the same temperature as the steam.

The equation that used to calculate latent heat can be given as:

$$H = ml$$

where H is the quantity of latent heat (J), m = mass of substance (kg) and l = specific latent heat for the change of state (J/kg). The specific latent heat may sometimes be called the 'specific enthalpy change'.

The following example shows how to calculate the total heat required to convert 3kg of water (m) at 60°C completely to steam at 100°C, given that the specific heat capacity of water is 4200J/kgK and the specific latent heat of steam is 2260kJ/kg.

Thus the specific heat capacity $H = mc\theta$
= 3 × 4,200 × 40
= 504,000J
= 504kJ,

specific latent heat $H = ml$
= 3 × 2,260
= 6,780kJ

Total heat = 504 + 6,780 = 7,536kJ

ENTHALPY

This may be described as the total heat content of a substance. For example, with reference to water, steam has a much higher total heat content than liquid water when they are both at 100°C. Steam at very high pressure and temperature carries a high enthalpy, which is useful as a means of transferring large amounts of energy. This is the reason why steam is used, as opposed to hot water, as a heating medium in many industries.

HEAT TRANSFER

Heat energy will tend to transfer from a substance with a high temperature to one with a lower one. A common human misconception is that of cold energy flowing, 'cold' does not in fact flow anywhere. Given bodies at different temperatures in the same environment, heat is exchanged between them until they are all at the same temperature. This transfer may occur in three ways: through conduction, convection or radiation. The process of evaporation can also transfer heat, which means that when latent heat is absorbed by a vapour in one particular place it may be released somewhere else.

Conduction

Conduction is the transfer of heat energy through a material without the molecules of the material changing their basic properties (McMullan, 2002). The process may occur in solids, liquids or gases, although the speed at which it occurs varies. As far as putting these three states of matter in rank order, then solids are the best, then liquids and finally gases. Different materials conduct heat at varying rates; metals, it will be appreciated, are good conductors, and poor conductors (often insulators) are materials such as plastics and porous substances that trap a lot of air and are used extensively in buildings. A good conductor requires a small temperature difference to transfer a given amount of heat per unit time; molecules vibrate rapidly and transfer heat as energy within the solid. Meanwhile, gases are poor conductors, and still air is a good insulator.

Convection

Convection is the transfer of heat energy through a material by the bodily movement of particles (McMullan, 2002). This process occurs in liquids and gases but not solids. The action occurs as, for example, air is heated and so expands; this expanded air is less dense than the surrounding air and the cool air displaces the heated air, causing it to rise. The new air is then heated, repeating the process and giving rise to a convection current. The stack effect is relevant here; this will be explained in Chapter 9.

Radiation

This is the transfer of heat by electromagnetic waves (McMullan, 2002). Heat may be transferred through space where conduction and convection are not possible; in fact, heat can travel in a vacuum, through air or in any other medium. The rate at which a body absorbs or emits radiant heat depends on the temperature and the nature of its surface. Rough surfaces, because they have a larger surface area, absorb or emit more heat than polished ones. Dark surfaces absorb more

heat and light; it is also true that good absorbers are good emitters and the opposite is also true. In a building example, heat radiation is discouraged by the use of aluminium foil insulation. The rate at which a body emits heat will increase with the temperature differential between the body and its surroundings. Indeed, every object is continuously absorbing and emitting heat from and to its surroundings.

RUMINANT AND NON-RUMINANT ANIMALS

Ruminant animals (such as dairy cows) have a high tolerance to temperature extremes, high or low, and their growth rate (or milk yield, if this is more applicable) is virtually unaffected by temperature. There seems little point therefore in controlling the environmental temperature of these animals. However, when these animals are in enclosed conditions it appears to be wise to provide a large volume of air with high ventilation rates. This then will effectively remove moisture and reduce the disease risk. In some circumstances fan ventilation (or forced ventilation) will be necessary to maintain a cool and acceptable atmosphere. Young ruminant animals, for example lambs or calves, are much more susceptible to adverse conditions and it is wise to design airflow patterns to minimize draughts at animal level; for example, high air speeds in cold conditions can cause severe problems with these animals.

Non-ruminant animals are much more sensitive in the way the environmental temperature affects their output. The two most important species in this category are poultry and pigs.

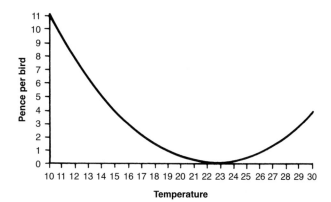

The costs to the farmer of incorrect temperature can be calculated by computer modelling. (Farm Electric, 1990)

Poultry

The LCT (lower critical temperature) and the UCT (upper critical temperature) are less well defined for poultry than they are for pigs. There are several factors that obscure any clear-cut critical temperatures: feathering, the strain of bird, the stocking density conditions and the feed requirement. Clearly these factors have a lot in common with those discussed below for pigs, but the recommendations are set out in a more general way.

Chick Environment
With chicks housed on the floor, the correct temperature may be ascertained by observing the behaviour of the flock, particularly when warmed by radiant heaters. If the birds move freely in and out of heated areas when drinking and feeding, then the overall temperature is acceptable. If the birds huddle and are reluctant to move from the groups, then it is likely to be too cold. If the birds do not form proper groups, spread their wings and show mouth opening, then the environment is likely to be too warm (see diagram on p.88). It is possible, as with pigs, to provide local-ized high-temperature 'brooding' areas under a radiant heater with lower house temperatures. Usually if the brooder area

temperatures are adequate, then the sur-rounding building temperatures can be taken to as low as 21°C without any prob-lems. The chicks may be attracted to high temperature areas in the house by the use of higher lighting levels in those areas. Another solution to heating during the brooding stage is segregation by polyethyl-ene curtains and the heating of that part of the building only. This can result in reduced heating costs for the farmer.

Layers and the Growing Bird
For layers and finishing meat birds over 3 weeks old, the Department for Environment, Food and Rural Affairs (DEFRA) have produced temperature rec-ommendations for given species. The results of incorrect temperatures are cal-culated from computer modelling, which will take into account the current prices of feed, eggs, and poultry meat. The figure above shows the 'cost of being wrong' with regard to heating.

Pigs

The factors that affect the LCT and the UCT may be summarized as:

• body weight and age; generally, as the

pig becomes older and heavier, its tolerance to low temperature increases and its tolerance to high temperature decreases
- feed level; feed levels vary according to the type of feeding system and the market for the finished pig; the effect of higher feed levels is to lower both the LCT and the UCT
- flooring; since a pig will spend a significant amount of time lying down, the insulating properties of the floor have an effect on the temperature tolerance of the pig, hence straw-bedded pigs tend to perform better at lower temperatures when compared with those pigs on slatted or solid floors
- air speed; draughts affect both the LCT and the UCT – this is generally taken as air moving in excess of 0.15–0.25m/sec; a well-designed ventilation system will subject the pigs to low air speeds in winter (to avoid chilling) and higher speeds in summer to give greater evaporative cooling and so raising the UCT; this situation is most easily achieved with fan ventilation systems; the LCT for pigs rises as the air speed increases and so draughts must be avoided in order to maintain their body temperature; in warm conditions an increased air speed can improve the daily weight gain and the feed conversion ratio
- radiant effects; at any given temperature the heat loss of the pig is affected not only by convection but also by radiation, for example, the radiative loss from the pig to a building will be high if the inside surface temperature of the surrounding is low, this will result in a high LCT that may be considered detrimental; on the other hand, in well-insulated structures the temperature difference between the inside surface of the building and the pig is not so high and therefore the heat loss of the animal is lower; this then gives a lower LCT; note that heating systems relying on radiant output to a large extent can have an important effect on the LCT
- stocking density; pigs have a lower LCT when housed in groups than when housed singly, thus their thermal interaction assists in keeping them warm; their UCT is also affected by group size' that is, the ability of pigs to lie clear of each other affects their capacity to dissipate heat; as a guide, 10 × 50kg pigs continuously produce 1kW of heat
- dunging habits; the temperature and the air distribution in a building have a significant effect on the dunging habits of the pig, the dunging area tends to equate to the cooler end of the pen, thus in partly slatted pen designs it is useful to induce, by careful direction of the incoming air, a cooler area above the slats to encourage dunging.

(These notes are based, with permission, on a summary taken from *Controlled Environments for Livestock*, by Farm Electric [1990].)

CHAPTER 8

Heat Losses from Buildings

Heat losses from buildings occur in a number of ways, the most common (in no particular order) being:

- insulation of building
- area of external shell
- temperature difference
- air change rate
- exposure to climate
- efficiency of services
- use of the building (McMullan, 2002).

These factors are discussed individually below.

TYPES OF INSULATION

Insulation saves money, and around the farm there are many benefits of it although they might not always be obvious. Insulation reduces the effects of the weather on the inside temperatures and there are many advantages of doing this. For example:

- internal temperatures more stable and easier to control
- lower fuel costs
- lower livestock feed costs
- less expensive heating or cooling equipment
- protection from frost
- less condensation
- less solar heat build up (ADAS, 1983).

Most buildings used in agriculture may be put into one of two groups for the purposes of considering the optimum thickness of insulation: those that use heat only from stored produce or livestock and have no supplementary heating, and those that need some form of artificial heating or cooling, that is, they use conventional fuels to maintain a required temperature.

Buildings that Require No Conventional Heating

Many buildings are designed so that any heat gained from crops or livestock is sufficient to keep the required temperature. This technique is commonly practised in pig-fattening houses. This results in a saving in the capital cost of heating equipment and the resulting fuel savings which make the system worthwhile. What happens is the building is made so that heat losses equal the heat given off by the stock, and careful design of ventilation and insulation make this feasible. In theory, there is no advantage in putting more insulation on the building than this calculated level. There is no financial benefit in attempting to keep a building warmer than its optimum temperature. However, in practice there is usually a safety factor designed into the system. It is wise to remember that it is always cheaper to put in adequate insulation at the design stage rather than have to add more later.

Buildings that Require Supplementary Heating

The factors that determine the most economical thickness of insulation are the capital cost of the insulation and the cost of the heat loss, that is, the cost of the fuel required to make up the loss. Any factor that has an influence on the costs of insulation and the fuel used will affect the thickness of the insulation. Therefore the factors that affect the optimum thickness may include:

- cost of insulation material
- cost of labour for installation
- expected life of the insulation
- anticipated future fuel cost
- design temperature of the building
- efficiency of the temperature control of heating equipment
- how often and for how long the building is heated
- local weather conditions and building exposure
- financial factors affecting interest rates (ADAS, 1983).

Some farm buildings require heat 24hr a day and 365 days per year and are therefore quite difficult to control. It is usually the case that the more a building is heated, the more it pays to insulate. There are many farm buildings which it is cost effective to insulate.

The Selection of Insulating Materials

Care and consideration must also be given to factors such as physical strength,

cost, flame spread, resistance to pests and condensation and moisture resistance (ADAS, 1983). Thermal insulators used in buildings are made from a wide variety of materials and are marketed under several trade names. They may, for instance, be classified by form under the following headings:

- rigid preformed materials (example: aerated concrete blocks)
- flexible materials (fibreglass quilts)
- loose fill materials (expanded polystyrene granules)
- materials formed on site (foamed polyurethane)
- reflective materials (aluminium foil) (McMullan, 2002).

Condensation and Moisture Resistance

When buildings are to be used for livestock, which produce large quantities of moisture in the air, this topic may be worth discussing in more detail. Surface condensation may be caused by warm, moist air within a building cooling below its dew point when it touches a cold structure (the dew point is the temperature to which moist air must be cooled for condensation of water vapour to occur). Hence surface condensation is most likely to happen in buildings that have limited insulation, because insulation has the effect of raising the inside surface temperature. Typical conditions in which surface condensation may occur will be unheated livestock buildings, such as sheep and cattle houses. A solution to this problem may be to ventilate the building (see Chapter 9). If the amount of condensation is quite small then this will cause no concern, most building materials are able to absorb small amounts of moisture, and then allow this to evaporate when conditions change. Of course, with metal structures and other cladding materials that do not absorb moisture the conditions will be different.

Condensation can be a more serious problem in agriculture structures. It is difficult to trace and will often go completely unnoticed until damage has been done to the insulation material and the structure. Livestock and stored produce will give off moisture, resulting in an increase in water vapour within the building compared with the outside. This can lead to the water vapour passing through the structure from inside to outside; the moisture can then condense into water within the wall or roof material. Interstitial condensation is condensation occurring within a structure or material. This can cause dampness, damages the structure and reduces the effect of insulation. Surface condensation is the deposition of moisture when humid air meets a cold surface; this can normally be prevented if the U-value is 0.5W/sq mK or less.

Interstitial condensation can be prevented by putting in a vapour check on the warm side (usually the inside) of the insulation. The vapour is then turned back into the building where ventilation will remove it. Care is needed with joints between materials in order to make the structure as vapour-proof as possible. Note that since most materials will allow some moisture to penetrate, it is considered good practice to ventilate any voids in order to remove any small amounts of water.

The hygroscopic properties of some common building materials are given below:

Material	Percentage water absorption
polyurethane	1.2
PVC	1.7
styrofoam	0.8
polystyrene	2.1

(ADAS, 1983)

Hence, while it will be seen that all the materials quoted above are fairly resistant to moisture absorption, some are notably better than others.

Materials which may be used as vapour checks include: aluminium foil, reinforced foil, polythene sheeting, bitumen tar, bitumen/water emulsion and plastic paints (ADAS, 1983).

The figure below depicts a vapour-barrier check layout.

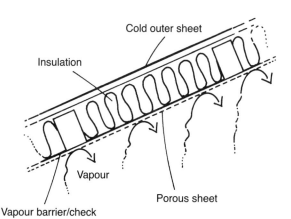

Diagram to show a vapour barrier construction. Insulation within the roof acts as a barrier to heat flow.

Some insulation materials resist moisture adequately themselves and can be used without a vapour check. A good example of this is extruded polystyrene which can be used in many agricultural applications without any additional vapour checks.

AREA OF SHELL

Quite simply, the greater the external surface area, the greater the heat loss from a building. The basic plan of the building can therefore have an enormous influence.

TEMPERATURE DIFFERENCE

The rate of heat loss through the shell is influenced by the temperature difference: that is, if the difference in temperature is large, there is a correspondingly large loss of heat.

AIR CHANGE RATE

Any warm air that leaves a building carries heat and is replaced by colder air. The warm air may simply leak out of the structure, or it may be controlled ventilation.

EXPOSURE TO CLIMATE

When a wind blows across a structure the rate of heat loss through it structure increases. This phenomenon is included in the standard value of external surface resistance when used in U-values.

EFFICIENCY OF SERVICES

Usually there is some wastage of heat as water or air is warmed. A good example of where this wastage is minimized is the recovery of latent heat in the flue gases as produced by gas burners. Any heat lost from a building, such as warm washing water being drained away, even though it is dirty, can be considered as waste.

USE OF BUILDING

Obviously, the number of hours per day, month or year that the building is in use has an effect on its energy consumption. If the building is to be heated through the night, such as a chicken rearing house, then the use of energy goes up. When different parts of the same building have different heating requirements, each section must be treated individually.

THERMAL CONDUCTIVITY (λ)

The thermal conductivity is a measure of the rate at which heat is conducted through a particular material under specified conditions (McMullan, 2002). This is calculated from W/mK, where W = watts, m = metre and K = degrees Kelvin. This measures the insulation properties of a particular material, the lower the λ-value, the less heat it will conduct and the better is the insulation. The table opposite shows λ values for a range of common construction materials.

Thermal conductivity (and density) of typical building materials

Material	Thermal conductivity (λ) (W/mk)	Thickness (mm)	R value (m²K/W)
Insulation			
Fibreglass/Mineral Wool			
Loose Quilt	0.04	100	2.50
	0.04	150	3.75
	0.04	200	5.00
Cavity Batts	0.038	50	1.32
	0.038	65	1.71
Timber Frame Batts	0.038	90	2.37
	0.038	140	3.68
Expanded Polystyrene	0.04	25	0.63
	0.04	50	1.25
Extruded Polystyrene	0.03	25	0.83
	0.03	50	1.67
Polyurethane	0.025	25	1.00
	0.025	38	1.52
	0.025	50	2.00
Vermiculite	0.075	100	1.33
Other materials			
Softwood	0.13	25	0.19
	0.13	50	0.38
Hardwood	0.18	25	0.14
	0.18	50	0.28
Plywood/Chipboard	0.13	9	0.07
	0.13	18	0.14
	0.13	22	0.17
Plasterboard	0.25	9.5	0.04
	0.25	12.5	0.05
add for Foil Backing			0.20
uPVC	0.4	2	0.01
Clay Brick	0.77	102	0.13

continued overleaf

Thermal conductivity (and density) of typical building materials *continued*

Material	Thermal conductivity (λ) (W/mk)	Thickness (mm)	R value (m²K/W)
Blocks			
Clinker Blocks	0.57	100	0.18
Aerated (Celcon)	0.18	100	0.56
Concrete Blocks	1.9	100	0.05
Granite	2.9	100	0.03
Limestone	1.7	100	0.06
Sandstone	2.3	100	0.04
Clay Tile	1.0	10	0.01
Concrete Tile	1.5	12	0.01
Roofing Felt	0.2	2	0.01
50mm Screed	0.4	50	0.13
65mm Screed	0.4	65	0.16
100mm Concrete Slab	1.3	100	0.08
150mm Concrete Slab	1.3	150	0.12
Render	0.57	12	0.02
Finish Plasters	0.57	3	0.01
Air gaps			
Cavity 25mm plus			0.20
10mm or less			0.10
Ventilated Loft Space			0.20
Surface resistance			
External			0.05
Internal			0.10

(Brinkley, 2002)

THERMAL TRANSMITTANCE (U-VALUE)

A U-value is a measure of the overall rate of heat transfer, by all mechanisms under standard conditions, through a particular section of a construction (McMullan, 2002); this is calculated from W/sq mK and is therefore the amount of heat that will pass through a 1sq m of the structure when the temperature difference between the inside and the outside is 1K. Improving the insulation means lowering the U-value. It will be seen that there is no gain in providing insulation in buildings where the inside and the outside temperature are quite close, unless it is to provide protection against frost. It therefore makes no practical economic sense to insulate cattle, sheep, many calf and some storage buildings. Some recommended U-values for farm buildings are given below.

piggeries where artificial heating is not normally required	0.6
piggeries where artificial heating is used (e.g., flat decks and farrowing houses)	0.5
poultry houses	0.5
frost protection only for crop storage buildings	1.15
solar heat gain protection for storage buildings	0.55
refrigerated stores at 2–3°C	0.45
refrigerated stores at 0°C	0.35
	(ADAS, 1983)

THERMAL RESISTANCE (R-VALUE)

Thermal resistance is a measure of the opposition to heat transfer offered by a particular component in a building element (McMullan, 2002); this is calculated

from sq mK/W. Thermal resistance is therefore a value obtained from the conductivity and the thickness of the material. It can be seen that the R-value is the reciprocal of the U-value, and therefore this should be as high as possible to prevent heat flow.

When calculating U-values therefore the following is true:

$$U = 1/R \text{ and also } U = \frac{1}{R_{si} + R_1 + R_2 \ldots} + R_{so}$$

where R_{si} = standard inside surface resistance, R_{so} = standard outside surface resistance and R_1, R_2 = resistance of that particular material.

For guidance, it is well to remember that there are always at least two surface resistances, that is, the inside and the outside surface. It is also wise to work to two significant figures only, since λ-values are no more accurate than this in the first place. The effect of timber joists or frames, wall ties, thin cavity closures, damp-proof membranes and other thin components may be ignored for the purposes of calculation.

EXAMPLES

1. (a) Calculate the U-value for a feed store with 210mm (8.3in) brick walls, with 50mm (2in) of foam.

   (b) Compare this U-value with that for a feed bulk store, having 200mm (8in) reinforced, aerated concrete panels.

   Data:
   λ values – brick = 0.84W/mK,
   spray-on polyurethane foam = 0.023,
   aerated concrete = 0.12,
   R_{out} = 0.060sq mK/W,
   R_{in} = 0.120
   U = 1/R

Solution:

(a)

	λ	r=1/λ	thickness (m)	R (= r × m)
brick	0.84	1.19	0.210	0.25
foam	0.023	43.48	0.050	2.174
out surface				0.060
in surface				0.120
				ΣR = 2.604

U=1/2.604 = 0.38W/sq mK

(b)

	λ	r	m	R
aerated concrete	0.12	8.333	0.200	1.667
out surface				0.060
in surface				0.120
				ΣR = 1.847

U = 1/1.847 = 0.54W/sq mK

Note: both U-values are below the recommended levels for frost protection (1.15W/sq mK) and protection against solar gain (0.55W/sq mK)

2. Calculate the composite U-value of the following situation:

10mm of rendering on the outside ($\lambda = 0.52$),
140mm of concrete block ($\lambda = 1.0$),
40mm of insulation ($\lambda = 0.029$),
3mm of cement board sheet ($\lambda = 0.4$) on the inside.
$R_{out} = 0.067$sq mK/W,
$R_{in} = 0.304$

Answer:
U-value = 0.52W/sq mK

3. A potato store needs a U-value of 0.55W/sq mK; a building is to be converted using styrofoam with a λ-value of 0.029W/mK; present U-values of the building are: roof 8.52W/sq mK, walls 3.64;

calculate the thickness of styrofoam required.

Answer:
U=1/R
(a) Walls

$$R(\text{exist}) = \frac{1}{U(\text{exist})} = \frac{1}{3.64}$$

$$U(\text{new}) = \frac{1}{R(\text{exist}) + R(\text{foam})}$$

$$0.55 = \frac{1}{R(\text{exist}) + R(\text{foam})}$$

$$0.55 = \frac{1}{(1/3.64) + R(\text{foam})}$$

$$0.55 = \frac{1}{(1/3.64) + m/0.029}$$

$$1/3.64 + m/0.029 = 1.82$$

$$m/0.029 = 1.82 - 0.27$$

$$m = 0.045m = 45mm$$

(b) Roof

$$R(\text{exist}) = \frac{1}{U(\text{exist})} = \frac{1}{8.52}$$

$$U(\text{new}) = \frac{1}{R(\text{exist}) + R(\text{foam})}$$

$$0.55 = \frac{1}{R(\text{exist}) + R(\text{foam})}$$

$$0.55 = \frac{1}{(1/8.52) + R(\text{foam})}$$

$$0.55 = \frac{1}{(1/8.52) + m/0.029}$$

$$1/8.52 + m/0.029 = 1.82$$

$$m/0.029 = 1.82 - 0.12 = 1.7$$

$$m = 0.049m = 49mm \text{ (or nearest thickness)}$$

4. A piggery requires a U-value of 0.5W/sq mK; a building is to be converted using styrofoam, $\lambda = 0.023$W/mK, present U-values are: walls 3.40W/sq mK, roof 7.50; what thickness of syrofoam is required on the walls and roof?

Answer:
roof 42.9mm, walls 39.2mm; in practice, this would be rounded up to whatever is convenient

5. Calculate the total heat loss from the building shown in the diagram below: (a) before insulation, (b) after adding insulation; what is the saving?

Data:
Inside temperature: 15°C,
outside temperature: 0°C,

U-values: walls 3.18W/sq mK dense concrete 203mm,
floor 1.13W/sq mK concrete on hardcore,
roof 1.70W/sq mK slates on felt,
door 0.18W/sq mK 25mm softwood,
windows (north) 5.00/sq mK single-glaze,
windows (south) 3.97 W/sq mK single-glaze
Roof to be lined with 100mm fibre glass $\lambda = 0.032$W/mK and 10mm gypsum plaster board $\lambda = 0.18$W/mK;
walls to be lined with 10mm plaster board with 25mm rigid urethane foam $\lambda = 0.022$W/mK;
windows will be replaced with double glazing with a 6mm air gap and U-value of north-facing 3.3W/sq mK and south-facing 2.67W/sq mK;
floor to be covered in 5mm, baked regranulated cork tiles $\lambda = 0.039$W/mK.

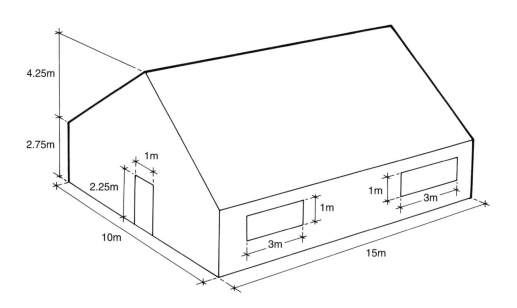

The dimensions of the building in question 5.

STRUCTURAL HEAT LOSS

Structural (or fabric) heat loss is caused by the transmission of heat through the walls, roof and floors. Provided that there are steady-state conditions, the rate of loss may be calculated by using the following:

$$Q = U \times A \times \Delta T$$

where Q is the rate of heat loss in watts, U is the U-value of the element under consideration in W/sq mK, A is the area of that element in sq m and ΔT is the temperature difference between the inside and the outside environment in °C; the total heat loss = $\sum Q$ for all surfaces.

Examples
1. A building has a total surface area of 850sq m, internal temperature is 17°C and the external temperature is -2°C, average U-value for the building is 0.62W/sq mK. What is the total heat loss from the building?

Answer:

$$Q = 0.62 \times 850 \times (17 + 2)$$
$$= 527 \times 19$$
$$= 10013W;$$
$$Q = 10kW.$$

2. A pig fattening house is 20m long, 5m wide, and 2m to the eaves, with a monopitch roof of 22 degrees. Given that the roof U-value is 0.72W/sq mK, the wall value is 0.69 and the floor value is 0.50, calculate the structural heat loss if the temperature difference between the inside and the outside is 12°. (It is usual in these circumstances to take a floor area of a 1m strip around the walls of the floor; for practical purposes, most of the heat loss is assumed to take place through this area; in other words, the heat loss in the middle of the floor is assumed to be negligible.)

Answer:
Area of the building = 304.3sq m,
heat loss of roof = 931.4W,
of walls = 1246.1W
and of floor = 276W.

HUMIDITY PROBLEMS IN BUILDINGS

The air humidity has a number of important influences since the amount of water vapour in the air controls the rate of evaporation of moisture from the external surfaces of the animal, especially from the lungs and respiratory tract (Sainsbury, 1988). In a ventilated building where air movement carries the water vapour away, continuous evaporation occurs, but the rate at which this happens depends on the rate of air movement and the percentage of saturation of the ambient air. Evidently, if the air moving in the building is already saturated, then no evaporation can occur.

The term relative humidity (RH) refers to the amount of moisture actually in the air compared with the amount it could contain at the same temperature, and expressed as a percentage (Sainsbury, 1988). Therefore the amount of moisture in the air at, say, 90 per cent RH, will depend on the air temperature. It is the actual amount of water vapour in the air that determines the amount of evaporation that occurs.

Sainsbury gives an example of what happens when a sample of air is breathed in and then out, and this will be summarized here:

- water vapour in air saturated at a freezing temperature of 0°C exerts a vapour pressure of 5mm Hg (mercury), as in winter conditions
- as air is inhaled into lungs, it meets moist mucous membranes at 37°C, with a vapour pressure of 45mm Hg
- the difference in pressure is 40mm Hg, therefore evaporation occurs quickly, saturating the air, now warmed to 37°C
- thus air entering at 0°C 100 per cent RH leaves at 37°C and also at 100 per cent RH, but with a vastly different moisture content
- as air is exhaled, it cools rapidly and surplus water vapour evaporates out
- this moisture is seen as a visible cloud, it also may condense and run on any cool surface.

The use of a psychrometric chart, such as the one shown in the next figure, can be used to find out the properties of air once some of the quantities are known. In our climate livestock will thrive adequately over a range of humidity values, from at least 30 to 90 per cent RH. If ambient temperatures are below the correct and ideal level, high humidities will exaggerate the cold stress because of the extra moisture surrounding the animal and taken into its lungs. If the temperature is higher than normal a high humidity will progressively reduce the animal's ability to remain cool by evaporation. If the air is very dry, say below 30 per cent RH, this may dehydrate the mucous membranes of the respiratory tract and make them more vulnerable to discomfort and

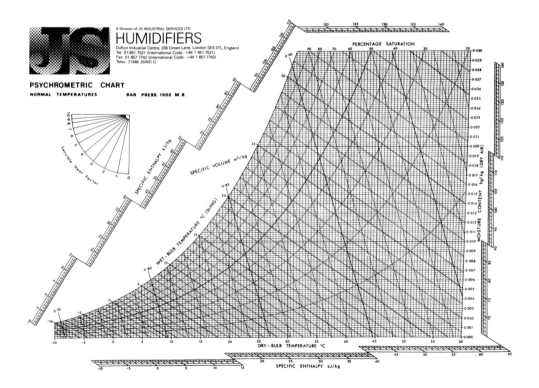

Psychrometric chart used in environmental design. (JS Industrial Services Ltd, 2005)

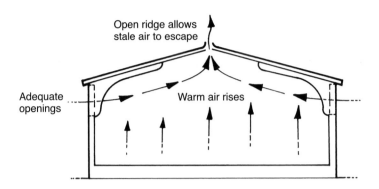

A good design of animal conditions. This shows the ideal air flow through a livestock building.

disease. It is therefore the aim of any livestock building to ventilate it so that any moisture produced by the animal is removed from the atmosphere as quickly as possible. In this way the animal remains in a healthy building and the structure itself will stay in an acceptable condition, see the figure above.

CHAPTER 9

Natural Ventilation

INTRODUCTION

Air movement or ventilation may be 'natural' or 'forced'. Natural ventilation refers to the way air will move through buildings by 'natural phenomena', whereas forced ventilation refers to a system that has fans and ducts that direct air forcibly wherever it is needed. Natural ventilation is now widely used for many types of livestock housing, including cattle, sheep, calf and certain pig buildings. Poultry and some pig houses rely on mechanical forced ventilation systems; these will be discussed in Chapter 11.

We first look at the natural currents of air around buildings, seeing how its behaviour may affect the conditions within. In a well ventilated building, which may feel cold to a human, there is good air movement that provides a better environment than a warm, stuffy one. Temperature may not be critical, but freedom from draughts certainly is. It is well to note that wet lying conditions can cause stock to lose heat and become more susceptible to disease due to the increased humidity. This may be reduced with good ventilation and air movement, acceptable stocking density and clean bedding.

In the United Kingdom the animals are unlikely to suffer from cold stress and they are housed to ease their management rather than to protect them from the environment. However, inadequate ventilation will cause animals a number of respiratory diseases that spread rapidly in damp conditions. Cattle and sheep in farm buildings are more likely to suffer from too little ventilation than from high air speeds. Calves are particularly susceptible to respiratory problems and require freedom from draughts, a dry bed and adequate air space.

Good ventilation in calf houses will remove the vapour produced by the calves and aid evaporation from the muck, bedding and any water spillages. It will also

The sloping side wall of a livestock house. Note the gaps or slots in the boarding and above the boarding, used for ventilation.

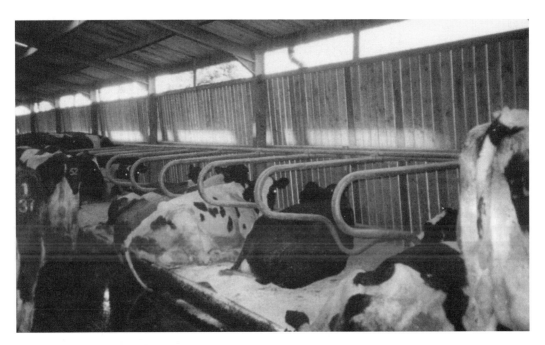

ABOVE: *Inside the livestock house in the preceding plate. Note the 'airy feel' to the environment.*

BELOW: *The inside roof space of a dairy building. Note the roof lights and the air gap at the ridge.*

Another dairy building, showing adequate space for the correct air flow. Note that the building is constructed primarily of timber.

minimize condensation and keep down micro-organisms which tend to thrive at higher humidities. Another useful advantage of adequate ventilation in a calf house is that it keeps dust levels down. The airspace should be 6m³ per newborn calf, rising to 12m³ at 12 weeks. Purpose-built accommodation with natural ventilation should have inlet areas allowing 0.05m² per calf and outlet areas of 0.04m² per calf (MDC, 1998). Converted buildings often need mechanical ventilation; this will be covered later.

The actual number of spans in the structure has a big influence on air movements and care needs to be taken to ensure each individual span gets sufficient ventilation. This may be ensured by stepping the building at the eaves connection or valley, or alternatively by using a breathing roof (see the figure on p.112 below). The roof pitch should not fall below 15 degrees since less than this will impair the airflow and the stack effect – see below. The surrounding area of the building is also important; closely spaced obstructions can have a deleterious effect.

A table showing stocking densities is given below, which should be used as minimal requirements and is calculated on the weight of animal at turnout.

Mass of Animal (kg)	Total Stocking Density* (sq m)
200	3.0
300	3.4
400	3.8
500	4.2
600	4.6
700	5.0

*Including bedded area and loafing/feeding area (taken from area allowances for cattle on straw-bedded yards (BS5502, 1991)

Any system of natural ventilation may be designed, and it is wise to pre-empt the worst conditions that may occur, that is, fully occupied on a still day. Draughts at stock level must be eliminated; this is particularly true for young stock such as calves. Air speeds should not exceed 0.25m/sec at the stock level. The building should not be overstocked because this often leads to ventilation problems. The ventilation openings, at top and bottom, must be adequate to do the job and must not be obstructed nor allowed to become 'overgrown' with rubbish. The total area of the inlet and outlet openings may be influenced by the weight and number of stock, the floor area of the building and height difference between the inlet and outlet openings.

The accompanying diagrams help to show how natural ventilation works in cattle or sheep houses and the sorts of building that do not give rise to good ventilation.

Diagram showing natural ventilation movement between the wall and cladding.

Diagram to show natural ventilation through spaced boarding of a livestock house.

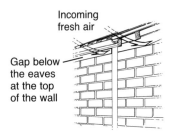

Diagram showing natural ventilation into a building via the gap below the eaves.

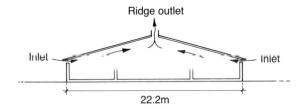

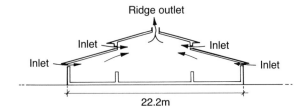

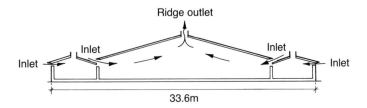

ABOVE: *Ventilation of multispan buildings. Diagrams showing how the span width affects free air movement within buildings and how the design may be used to assist.*

RIGHT: *Polypropylene or high-density polyethylene mesh used between the wall of a building and the eaves to aid ventilation.*

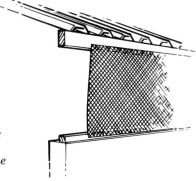

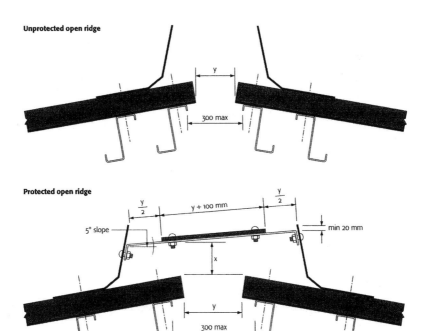

Open ridge design. An example of one manufacturer's design of ridge. (Marley Eternit, 2004)

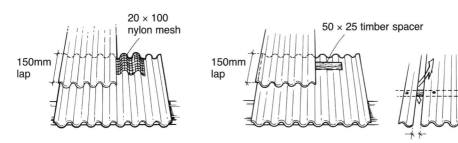

A breathing roof with nylon mesh. Here a strip of mesh assists in the air flow through the building. (Marley Eternit, 2004)

Two methods of achieving increased ventilation via (left) a breathing roof using extra battens and (right) a spaced roof. (Marley Eternit, 2004)

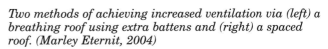

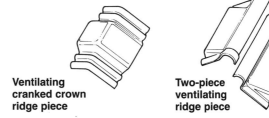

RIGHT: Types of prefabricated ridge fitting (Eternit). (Marley Eternit, 2004)

Ventilating cranked crown ridge piece

Two-piece ventilating ridge piece

There are two natural forces that push or propel air through a building: pressure and temperature.

THE PRESSURE EFFECT

There are two separate aspects to the pressure effect (or wind effect as it is sometimes known): aspiration (the effect of wind blowing over a building and sucking the air out (a diagram on p.117 shows how this works), and perflation (the through draught across a building created by the wind or breeze).

As wind blows on a roof the usual relationships of pressure can be shown: the wall under pressure has a positive value, while the other faces have a negative value, that is, a suction. This is true with roof pitches up to 35 degrees. However, with pitches of 45 degrees the windward side of the roof is pressurized. It is worth pointing out that as the roof and leeside wall are under suction, holes or openings in the roof or wall will act as outlets when the wind is blowing across the building. Buildings that are wide, in excess of 18.3m (60ft), are notoriously difficult to ventilate because the amount of energy to move the vast quantities of air may not be available. Air may not reach the central part of the building at all in some conditions, and the air that does may well pick up large amounts of moisture from the animals and bedding. Having outlets at the ridge and all over the roof slope can partially overcome this problem, but some areas may remain poorly ventilated.

There is always a need in any livestock house to avoid draughts. The effect of one is to remove heat from the surface of the animal and thus lower its temperature. It is exactly the same with human beings. Some type of baffling at the inlet to direct air away from the livestock is to be preferred and fittings in the house may affect the direction of air flow (purlins at the roof level may be advantageous in directing the air on to the stock if the air inlet is at eaves level). Generally speaking, the inlets to a livestock house should be situated above the level of the stock if high inlet speeds are expected. The outlet is most unlikely to be the cause of draughts in the livestock building.

THE TEMPERATURE EFFECT

This, also called the stack effect, refers to the way air is warmed by the livestock (it also becomes fouler), becomes lighter and rises as air heated by a radiator would, only to be replaced by cooler, fresher air from outside. In order to make this method work properly, there must be two openings in the building, normally 1.5–3m (5–10ft) apart, to act as the inlet and the outlet. The building span should generally be less than 22–26m (72–85ft) for ventilation to work properly. The diagram on p.108 shows how a building should be equipped for successful natural ventilation.

The figures below show both the wind and the stack effect inside a cattle house.

The stack effect is the principal method by which a good airflow is achieved in winter when the ambient temperature is low. The continuous cycle of air change throughout the building ensures that it can 'breathe' during still days. On the other hand, it is not as effective in summer with the weather warm and humid. When the ambient conditions do not readily give rise to the stack effect, the building must rely for its ventilation on perflation and aspiration.

Poorly ventilated buildings may be

ABOVE: *A cubicle house for dairy cows. Note the ventilation gaps in the side and ends.*

BELOW: *A different livestock house, with more ventilation (Yorkshire boarding) shown.*

ABOVE: *An open-fronted pig building constructed with steel, timber and fibre-cement on the roof.*

BELOW: *A modern pig building constructed simply with concrete blocks and a fibre-cement roof.*

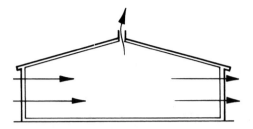

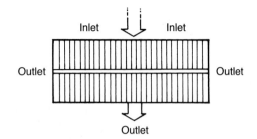

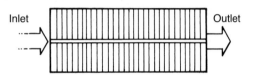

ABOVE: *The wind effect. A diagram to show how air flows through a building when it is subject to wind movement.*

RIGHT: *Wind effect. Two plan views of a building showing air movement when the wind is blowing from different directions.*

relatively easy to spot: they may smell of ammonia, have cobwebs hanging from the internal structure and a black, fungal growth on the underside of the roof. In order to get a clear picture of how the ventilation is working in a building, it may be necessary to carry out smoke testing. A simple device is available for this and, by closely looking at the rate and efficiency of the air's progress from the inlet to the outlet, air movement can readily be detected, although the air capacity cannot.

CALCULATING THE SIZE OF VENTILATION OPENING – CATTLE BUILDINGS

This method of calculation gives the minimum areas required for adequate, natural ventilation in still conditions. It applies to a ventilation system that has openings at the eaves level and an open ridge, that is, the normal arrangement for a cattle or sheep house. To follow an example it is best to put in some figures and so we shall imagine a building 24m (79ft) wide and 43.2m (142ft) long, with

a 15-degree roof pitch. It has a central feed passage and cows are housed in two groups of sixty-four in cubicles on either side of the feed passage. The average weight of the cows is 600kg (12cwt).

Step 1
Calculate the total floor area per animal (A), including feed passages:

$$= \frac{\text{total building area}}{\text{number of cattle}}$$
$$= 24 \times 43.2/128$$
$$= 8.1\text{sq m/animal}$$

Step 2
From graph (i) in the figure overleaf, using liveweight and total floor area per animal, find the outlet area, A^1.

A is 8.1sq m, liveweight is 600kg, A^1 = 0.14sq m/animal; this is the outlet area for a building with a height difference of 1m between the inlet and the outlet; this needs to be corrected for the actual height difference.

Step 3
Calculate the height difference, H; from the table overleaf a 15-degrees roof pitch has a rise of 1 in 3.7

(i) Graph to show the outlet area for height differences of 1m

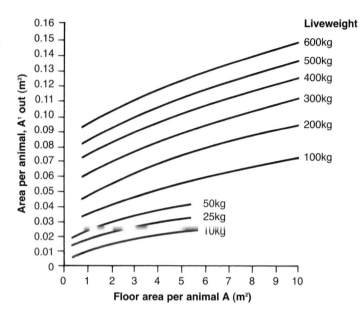

(ii) Figures for height factor

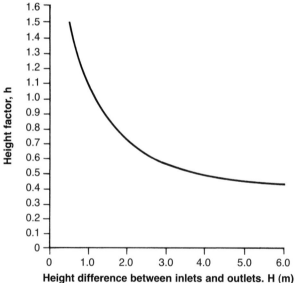

Cattle natural ventilation. (ADAS, 1992)

Relationship between roof pitch and rise					
Roof pitch in degrees	10	15	22.5	30	45
Rise	1 in 5.5	1 in 3.7	1 in 2.44	1 in 1.72	1 in 1

Height difference H

$= \dfrac{\text{½ span}}{3.7}$

$= 12/3.7$

$= 3.2\text{m}$ (that is, a ratio of 3.2 to 1).

Step 4
From graph (ii) in the figure opposite, find the height correction factor h; for a height difference of 3.2m, h = 0.56

Step 5
Multiply A^1 by h to give the corrected open ridge area per animal:

$A^1 \times h$ = actual open ridge area per animal:
$= 0.14 \times 0.56$
$= 0.078\text{sq m/animal}$

Step 6
Calculate the total area of open ridge.

Area of open ridge per animal
× number of animals:
$= 0.078 \times 128$
$= 9.98\text{sq m}$

Step 7
Calculate the width of open ridge:

$= \dfrac{\text{total area of open ridge}}{\text{length of building}}$

$= 9.98/43.2$

$= 0.231\text{m}$

In practice, this would be 250mm

Calculate inlet area:

inlet area = outlet area × 2

this could be provided by:
(a) continuous gap at the eaves: a gap of 250mm (the same width as the ridge) along both eaves, that is, × 2

(b) spaced boarding: if spaced boarding with a void area of 20 per cent were used (say 100mm boards with a 25mm gap) the depth of spaced boarding required at each eave is:

$\dfrac{\text{inlet area}}{\text{\% void area}}$

$= 0.25/0.20$

$= 1.25\text{m}$ (on each side)

CALCULATING SIZE OF VENTILATION OPENINGS – SHEEP BUILDINGS

A similar procedure might be used for estimating the size of ventilation openings on sheep buildings as that used for cattle sheds. For example, a building measuring 27.5m × 9m houses 180 sheep, each weighing 30kg; the pitch angle of the roof is 22.5 degrees. To calculate the size of open ridge and inlets required, the following procedure may be adopted.

floor area per animal, A
$= \dfrac{27.5 \times 9}{180}$
$= 1.38\text{sq m}$

From graph (i) in the figure overleaf, estimate A^1 out = 0.02sq m²/animal, this gives the outlet area for a height difference of 1m; but this has to be corrected for actual height difference:

height difference, H
$= \dfrac{\text{½ span}}{\text{rise}}$
$= 4.5/2.44$
$= 1.84\text{m}$

(the rise is taken from the table on p.118)

(i) Graph to show the outlet area for height differences of 1m

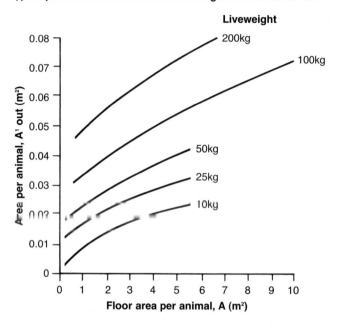

(ii) Figures for height factor

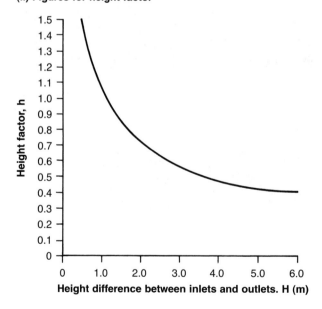

Natural ventilation of sheep buildings. (ADAS, 1992)

Actual height factor, h = 0.74 (taken from graph (ii) opposite); multiply h by open ridge area per animal

$$= 0.74 \times 0.02$$
$$= 0.015\text{sq m/animal}$$

thus total open ridge area

$$= 0.015 \times 180$$
$$= 2.66\text{sq m}$$

and width of opening

$$= 2.66/27.5$$
$$= 0.0967\text{m}$$

In practice, this would be 100mm.

The inlet area = outlet area × 2, for instance, with spaced boarding down both sides and void area of 15 per cent, then depth of boarding

$$= 100/0.15$$
$$= 667\text{mm}$$
$$= 0.7\text{m}.$$

Alternatively, a continuous opening may be specified down each side of 100mm (Methods based on *Planning Sheep Housing and Planning Dairy Units*, 1992, FRBC).

ABOVE: *A relatively old pig building showing small pens and automatic feeding outlets to each pen.*

RIGHT: *Another view within the same building as in the preceding plate.*

ABOVE: *A photograph of an old dutch barn converted and successfully used for pig housing.*

LEFT: *This plate shows a converted pig pen situated within an old farm building.*

OTHER FACTORS AFFECTING NATURAL VENTILATION

Radiation on the Roof

High temperatures at roof level may be created by the sun, and this can lead to variations in temperature and humidity across the building. For example, high humidities can occur on the north side of an east–west house. Perhaps having an asymmetric east–west building can help to minimize this problem. Sometimes re-radiation from the roof can lead to stock discomfort, and the insulating of the roof may be an answer.

Effect of the Housing System

Where animals are free to move around, this will aid the mixing of the ventilating air. Local air speeds of 3.2–4km/hr will be created. Solid pen divisions may lead to stagnant air pockets; draught-free conditions may be required for young stock such as calves, and therefore solid pen divisions may help if they are situated across the air flow.

Wide Spans

Many wide span buildings have a minimal height at the eaves and therefore have a large, dead, air space in the roof. They have a larger total volume of air per unit of floor space and this leads to high inertia of air inside the building which requires more energy to move it. In wind-effect conditions this means that some parts of the building will remain unventilated. Natural ventilation seems to work well in buildings of up to 18.3m (60ft)

A pig building with the dry pellet feed hopper outside. Notice that the building has plenty of ventilation above the dwarf wall.

span widths; beyond this there could be problems. Often multispan buildings tend to work better as far as air is concerned. The air volume in the building is less and there will be more than one high point present. Each separate span has its own ridge outlet, but problems can arise in wide span buildings owing to valley gutters and the difficulties these create in terms of getting air into the buildings between the spans. One solution to this might be to have variations in the height of buildings.

Gable Ends

These should be provided where sufficient side inlets or outlets are not possible. Gable end openings operate by wind effect but may not help the stack effect by not allowing sufficient height between the inlet and outlet. The gable ends will act as either inlets or outlets according to the direction of the external wind (see diagram on p.117). Often buildings have small areas in the gable ends that can be left open e.g. those above fully sheeted doors.

Very often in windy conditions, undesirable draughts can occur at the corners of buildings owing to turbulence. These may be overcome by fully sheeting the first 1.5m (5ft) from the corner of both the side wall and the gable ends. The use of solid gates is preferable to ordinary railed gates and can help to reduce draughts, especially at the ends of the building. The end pens of buildings must be considered separately in order to reduce draughts for the livestock.

Rain and Snow

In general terms, there should be a requirement to prevent the entry of rain and snow. This may mean the fitting of a ridge cap and the correct sizing of inlets to prevent rain from entering. Powdered snow might be blown into any naturally ventilated building; too large a gap will lead to potential snow problems, while too small a gap will create insufficient ventilation. A compromise must be sought, but all the time it is well to remember that livestock are used to conditions far harsher than humans can tolerate. In high winds all parts of a building with a roof pitch of less than 35 degrees (except the windward side) will be experiencing suction. Light airborne particles of snow are unlikely to enter the roof and will remain airborne to fall to the ground beyond the building. Meanwhile, the windward wall can give problems. Sloped upstands can improve the ability of the ridge to keep out snow and rain, and this is a design favoured by some manufacturers.

Capped Ridge Outlets

Ridge caps are designed to keep out inclement weather, although perhaps there is a tendency to do away with these on some new buildings. Chimney outlets can be good natural ventilation outlets, not only do they increase the stack height in the building but they can also be designed to prevent snow or rain from entering. Often these raised ridge outlets are found on calf houses or in piggeries.

METHODS OF ACHIEVING NATURAL VENTILATION

There are several methods that are used on buildings in order to achieve natural ventilation flow. The most common are:

Wall Cladding

- fixed space boarding: pressure-treated boarding is used as a cladding but with a gap left between each board, it is often called Yorkshire boarding; the boards should be fixed with galvanized nails and the length of the boards determined by calculation, such as that illustrated above; it gives a pleasing exterior finish to the building (see the figure on p.112 [centre left])
- adjustable spaced boarding: this is a version of the previous method but using fixed outer slats and sliding inner slats; with this version adjustments may be made to the sliding slats, depending on the weather conditions prevailing
- louvred, plastic-coated, sheeting/polypropylene or high-density polyethylene mesh; this may be used as a substitute for spaced boarding, 18-per cent openings are normally supplied with louvred sheeting, otherwise a plastic mesh may be used to act as a windbreak; 45-per cent openings are usually achieved
- slotted hardboard: exterior grade hardboard with 25mm × 5mm slots can be placed horizontally or vertically and give 30 per cent openings; as with all these materials, the light transmission is reasonably good and the draught reduction is acceptable
- protected inlet: by building the wall inside the stanchion and extending the outer cladding, sufficient area for ventilation can be incorporated, again, draughts are reduced (see the figure on p.112 [top])
- adjustable, hinged flap: this allows air to be directed into the building; the inlet area is determined by the size of flap and the amount of tilt of the flap; the materials usually used to make the flaps are exterior grade hardboard on a

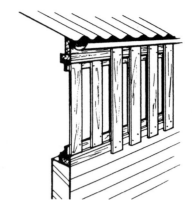

Adjustable, spaced boarding that may be used on the walls of livestock housing.

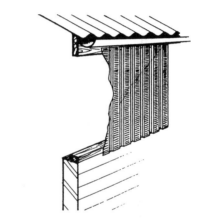

Louvred, plastic-coated sheeting that may be applicable for livestock housing.

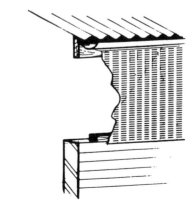

Slotted hardboard that may be used for livestock housing ventilation.

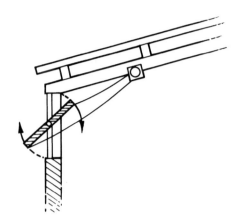

Adjustable, hinged flap that could be applicable for the sides of some livestock buildings.

timber frame or plyboard and adjustment will be by nylon rope with a pulley system; this method is normally used in pig and chicken houses.

Roofing

- venturi ridge: an apron flashing made from fibre cement or plastic-coated steel flashing may be used as a venturi effect (as previously mentioned); the area of outlet is determined by calculation as completed above
- capped ridge outlet: often in plastic-coated steel sheeting (or fibre cement), this has large, separated ridge outlets, capped by a pressed sheet and is supported on a length of steel flashing (see the figure on p.113 [top])
- slotted corrugated roofing: the idea of this roofing sheet is to provide a 'breathing' roof; 1.4 per cent openings are usually provided; similarly, upturned, corrugated sheeting may be used with a gap between each sheet, this forms a venturi slot between each sheet, and typically a 25mm gap would give 3.7 per cent openings; conventional roof sheets may be raised by using a batten or washers to allow air to circulate underneath, a batten of treated timber of 20–45mm is suitable; these last methods may often be done as a DIY approach on an existing building in order to attempt to improve the ventilation (see the figures on p.113 [centre left and right]).

CHAPTER 10

Automatically Controlled Natural Ventilation

INTRODUCTION

The Scottish Farm Buildings Investigation Unit (as it was then called) developed a system for pigs that was an extension of natural ventilation and designed to give good temperature control at low running costs. The system was made to monitor the inside temperature of a pig building and then automatically alter the ventilation rate whenever it was required. Therefore the system was used to optimize the performance and the productivity of the pigs inside the house by keeping the internal air temperature at the ideal level. From previous information about the upper and the lower critical temperature (UCT and LCT), a ventilation system should maintain the internal temperature to the thermoneutral zone, despite the conditions outside the building. It normally has the temperature in the building to within ±2°C of that set on the control thermostat, which is 2 or 3° (above the LCT (SFBIU, [n.d.]).

HOW AUTOMATICALLY CONTROLLED NATURAL VENTILATION (ACNV) WORKS

The system comprises a controller unit, a thermostat, adjustable flaps in the side walls of the building and a motor or linear actuator to operate the flap. The layout of a typical system is shown below. The thermostat measures the internal air temperature at intervals (which are preset) and relays these to the

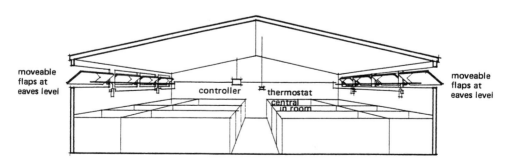

Layout and components of an ACNV system. (SAC [n.d.])

controller. If the temperature is above or below that required, then the controller switches power to a mechanism that either opens or closes the flaps in order to adjust the ventilation rates. In other words, if the house is too hot the flaps are opened, if it is too cold they are closed.

The controller has two timers and a switching mechanism. A control or master timer determines the sample time or the length of time between testing the house temperature. The other timer determines the time for which power is supplied to move the flaps, hence this adjusts the ventilation rate at each sampling. The stability of the whole system depends on how the sample and run time are set, and typically the sampling time may be adjusted to 3–10min – a time of 5min has been found to be adequate. Indeed, if sam-

pling is too frequent, the system may become unstable. Run time is set between 2 and 5sec, depending on the number of stages between being fully closed and fully open.

It is essential to:

* position the controller in an accessible position but in one which avoids damage
* use a controller which has both automatic and manual operation
* ensure the correct setting of timers for system stability
* use a controller which is robust and easily maintained (SFBIU, [n.d.]).

The flaps are normally positioned at the eaves level on both sides of the building, and these act as both inlet and outlet. They are opened in small steps, having

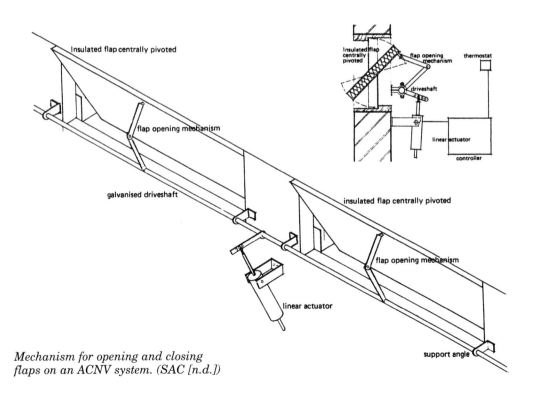

Mechanism for opening and closing flaps on an ACNV system. (SAC [n.d.])

around fifteen between the fully open position and the fully closed state. If the house temperature is correct, the flaps do not move and can remain still for hours, particularly in the summer. The flaps normally open in the morning in a series of movements which may take 50min or so; during the day they can stay fairly steady, depending on conditions, and then close in a similar manner in the evening as it cools. It is important that the flaps seal all round to avoid air leakage and moisture penetration. It is also essential that the flaps are close fitting, robust and easily moved. Plywood has been tried for these flaps, and, although it is cheap, it does warp and cause poor fitting. Condensation problems may also occur in the winter with these flaps. Bonded, insulated panels are much better, both in overcoming condensation problems and in being light and rigid; they also have an inherent stiffness that helps to avoid warping problems. An alternative material which is sometimes used is clear, double-skin polycarbonate. Obviously this allows natural light to enter the piggery and thus create a better environment for the stockman. Stiffening rods, usually surface mounted and running the full length, are used with these flaps.

The thermostat that monitors the house temperature has three positions: higher than set temperature, within the range of set temperature, and below set temperature. If the temperature sensed is above the set range, the flaps will be opened by one step unless they are already fully opened. If the temperature is below the set range, the flaps are closed by one step unless they are fully closed already. Of course, if the temperature detected by the thermostat is within the set band, the flaps stay still. Typically, the 'do nothing' band will be ±2°C of the required temperature. The temperature set on the thermostat is normally 2 or 3°C above the LCT of the pigs. This LCT will vary according to the particular circumstances in which the pigs are housed, and the thermostat usually has a range of 10 to 40°C.

There is a potential problem if the power fails since this could put the livestock at risk. Therefore an alarm is fitted to alert staff in the event of failure. It must operate separately from the control system itself, be independent of the mains electrical power and have an in-built test facility. This will warn staff of a problem when the temperature goes above or below the pre-set limits due to a mains failure or whatever; a member of staff has then simply to disconnect the arm of the drive shaft to alter the position of the flaps by hand.

ADVANTAGES AND DISADVANTAGES OF ACNV

Advantages

- ACNV is a low-energy system and consequently has low running costs; this will make the system increasingly attractive
- ACNV gives good temperature control, particularly at low ventilation rates
- ACNV will operate with low air speeds within the building, even in windy conditions
- ACNV does not create dusty conditions
- ACNV is quiet in operation; normally a fan system is noisy, an ACNV system clicks as it switches on and runs for a few seconds
- based on a combination of the stack effect and wind, ACNV does not depend on forced air movement to function so

there is less likelihood of major problems should the power fail
- ACNV is virtually fail-safe; if there is a
power failure when the flaps are open
they would remain in that position and
ventilation would be provided; power
failure warning devices can easily be
included in the system if desired.

Disadvantage

- ACNV will not give any control of air
direction within the building (based on
SFBIU, [n.d.]).

DETERMINING THE FLAP SIZES

The easiest way to explain this is to take
an example. Say a system is to be
designed for a 400-place, fully slatted
fattening house, 36m (118ft) long and
11m (36ft) wide. The pigs are going in at
30kg (79lb) and finishing at 90kg (200lb).
The farm is in Essex. What are the appropriate ventilation flap sizes?

Step 1
From the next figure, the summer design
temperature is 23°C.

Step 2
The average live weight of pigs is 60kg.

Step 3
Read off from the graph in the next figure
for 60kg pigs the flap area per pig for 23°C
on the curve for slatted floors; the answer
is around 0.026sq m (0.28sq.ft).

Step 4
Multiply this figure above by the number
of pigs in the house:

*A map of Britain showing approximate
design temperatures (°C) applicable for
ACNV design. (SAC [n.d.])*

$$= 400 \times 0.026$$
$$= 10.4\text{sq m (112sq.ft)}$$

This area is required per side wall.

Step 5
Flaps are chosen to fit the structure of
building, thus they might be 1.2m x 0.5m
(4ft x 1.6ft); therefore 18 flaps would be
required along each wall (1.2m x 0.5m =
0.6sq m and 0.6sq m x 18 = 10.8sq m).
(Interpolation between the graphs is
allowed, making the calculation easier for
all situations.)

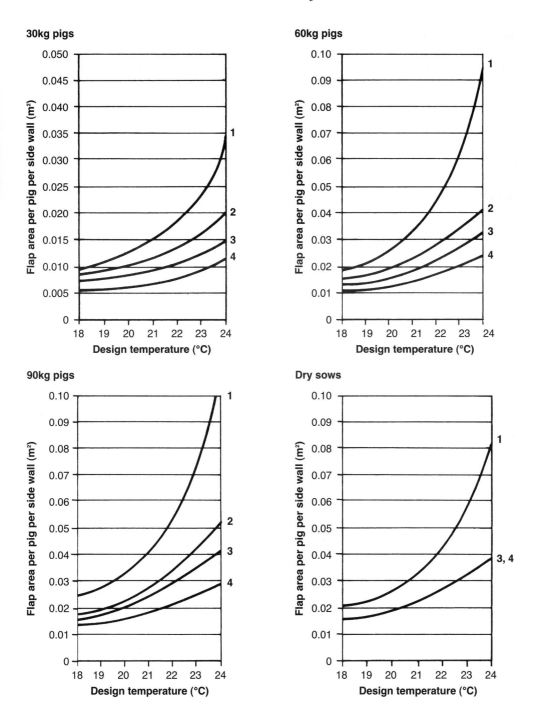

Key
1 Bedded floor   **2** Perforated metal floor   **3** Concrete slatted floor   **4** Solid concrete floor

Four graphs used for sizing ACNV flaps. (SAC [n.d.])

A large poultry unit in a building which houses around 25,000 chicks, of varying age. The house has fully automatic ventilation, heating, feeding and watering devices fitted and is also installed with systems that control lighting and humidity.

The roof of the same building. Note the ventilation openings at the ridge and eaves.

CHAPTER 11

Forced Ventilation Systems

INTRODUCTION

It is vitally important to have the correct amount of ventilation air circulating around the animals while they are kept inside. If this is not supplied in a system called 'natural ventilation', as previously described, then it will come from one or more fans, forcing the air around the building in a prescribed manner. Ventilation systems may affect some or all of the following:

- moisture level
- air temperature
- moisture condensation on surfaces
- uniformity of air temperature
- air speed across animals
- odour and gas concentration
- disease organism level and combustion fumes present.

As Carr puts it, the ventilation system in any (pig) house has three functions:

- to provide a micro-environment which is suitable for the pig living in that environment; this is determined by satisfying its known welfare requirements and maximizing profitability per cubic metre of housing
- to provide air quality which is suitable for the health and safety of the people working in the building
- to protect the building and equipment from corrosion (Carr, 1994).

ASPECTS OF DESIGN

The fan pattern normally used is the propeller type. This is relatively cheap and easy to maintain and will move large amounts of air at low back pressure – certainly when compared with other types such as an axial flow model. More discussion of different types of fan will follow in Chapter 13.

There are several factors affecting back pressure and fan efficiency (based on the Farmelectric Centre, 1990):

- size of inlets and outlets; air speed through inlets and outlets should not exceed 2.5m/sec (8ft/sec) (at maximum fan speed setting); this figure is acceptable in winter, whereas in summer a figure of at least five or six times higher is required
- internal construction of ducts, inlets, cowls: it is advisable that the internal surface of the cowl should be as smooth as possible, that is, all construction battens and such like should be placed on the outside away from the air flow
- correct housing around fan impeller; badly made and ill-fitting housings reduce the airflow considerably; the normal arrangement is for fans to have a proprietary ring diaphragm for mounting or a 'bell mouth' which provides for smooth entry/exit for the air
- performance at reduced speed; the throughput of air is directly

proportional to fan speed, thus reducing the speed by 50 per cent reduces the output by 50 per cent; on the other hand, the pressure developed by the fan (and hence the backpressure against which the fan can operate) is proportional to the square of the fan speed, hence halving the fan speed reduces the pressure developed by 75 per cent, this can be significant when one studies the whole ventilation system that may be in a piggery or poultry house.

It is well to note that if fans are blowing air into a building (pressurizing), the fan aperture acts as an inlet. If, on the other hand, the fans are pulling air out of a building (depressurizing), the fan aperture is the outlet. Different systems may use one method as opposed to the other for various reasons; this will be covered

LEFT: A newer poultry building on the same farm as in the preceding plates. It uses metal cladding, with ridge vents.

BELOW: A view from the other side of the same building. Note the two dry meal feeders used for the birds in the house.

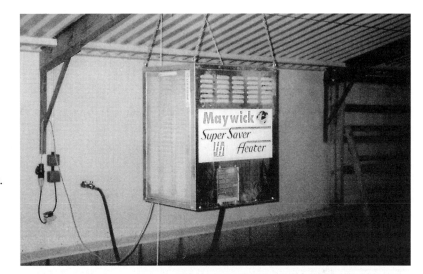

Within the same building as in the two previous plates. The gas-powered warm air heater (one of many) provides sufficient warmth for the birds.

A different view from within the same building, showing the chicks and mechanized feeders.

later. Air speeds in a building around an outlet reduce in proportion to the cube of the distance from that outlet. High air speeds therefore occur only near the outlet aperture, so a designer need not worry about draughts when he is choosing a position for the outlet.

Air passing through an inlet into the building tends to 'squirt' through and, as a consequence, can be highly directional. This means that it can be felt far away from the inlet aperture and can be the cause of draughts. Sometimes it can be used to advantage when a relatively high air speed is required. In conclusion, the inlet must be positioned with some care with regard to its effect on an animal's environment.

The temperature and the speed of the incoming air relative to the air already in the building has a marked effect on the temperature gradients and circulating air

Summer

Winter

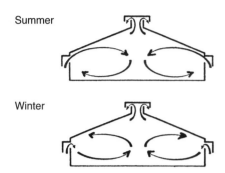

Air patterns in summer and winter conditions. (Farm Electric, 1990)

currents in the building. For example, cold air entering at low speed drops just beyond the inlet; air entering at high speed can travel a good distance at high level before mixing with the building air. These two situations can occur in the same building, one in the winter and the other in the summer. In some pig houses, for example, this might lead to a reversal in the dunging and lying area, resulting in animals which become dirty, with a reduced feed conversion ratio which becomes expensive.

WIND INTERFERENCE

Wind protection is an important but often forgotten part of the design of ventilation systems. However well the control system for the piggery, say, is designed, if the inlets and the outlets are unprotected there will be large fluctuations in temperature. The pigs will suffer from too much or too little fresh air and from poor temperature control, and probably, most important of all, from cold draughts.

The reason for protecting the inlets and outlets becomes apparent when one considers wind pressures. The wind blowing against a vent produces a pressure that is proportional to its velocity. As one might expect, the pressure is greatest when the wind is at right angles to the wall and therefore blowing direct against a vent. Rarely, of course, are wind speed and direction constant, and therefore the wind velocity relative to a wall absolutely constant. Normally, the stronger the wind the more it varies in direction. However, when wind hits the wall of a building it behaves like a jet of water and sprays out in all directions, swirling and eddying so that, even if the inlet or the outlet is not facing the wind direct, it can still be affected by the wind turbulence over the whole building.

Let us consider an exhaust propeller fan exposed to an opposing wind. This fan will work against a resistance that is produced by the pressure of the wind. Taking the example of a 450mm (18in), 960rpm propeller fan, this will give approximately 1cu.m/sec under free airflow conditions. This will be reduced to 0.6cu.m/sec by a resistance of 5mm water-gauge, and at a resistance of 7.5mm water-gauge it will give practically no air at all. Therefore a 'fresh breeze' blowing towards the fan direct will totally inhibit its function.

The wind will not always reduce the output of the exhaust fan considered. If the wind hits the wall at an oblique angle it may produce a sucking effect (called induction). Maximum induction will occur if the wind is substantially parallel to the wall. Of course, in this example an outlet or exhaust is considered. The situation would be the other way round in the case of an inlet. Maximum airflow will occur when the wind is at right angles to the wall and minimum when it is parallel to it.

Ventilation Chimneys

There are a number of designs of ventilation chimney, but the type shown in

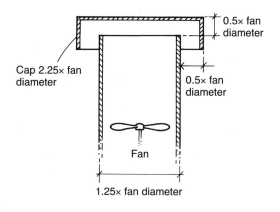

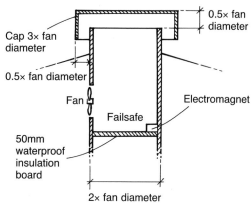

Diagram showing a fan chimney that could be suitable for a pig/poultry house.

Diagram to show typical dimensions for a fan chimney suitable for a pig/poultry house.

the figure above left is typical of a well-protected exhaust fan chimney. This maintains full air delivery, irrespective of the wind direction by virtue of the four-way outlet. Though the output from one side may be reduced by wind pressure, the same wind will assist the flow from the others by induction. A good design will maintain about 95 per cent free airflow delivery.

One variation of the fan chimney for a house having a pressurized ceiling (see below for system descriptions) is to have a fan on the side of the chimney on a horizontal axis rather than in the centre of a chimney on a vertical axis, see figure above right. The advantage of this design is that if one has a fail-safe, drop-out panel and there is a power failure, then the chimney is unimpeded by any fan. In this case also a larger size of chimney is used.

Outlets

The object of the baffling is to try and make the outlets as independent as possi-

ble of the wind. It is important to ensure that air is expected to go through one or two right angles and that it is not restricted by the cross-sectional area of the vent that it has to traverse. Three examples of the large number of such baffles in use are shown in the three figures overleaf (top and centre).

Inlets

The same considerations must be taken into account as for outlets, except that in this case care also has to be taken that any inlet design does not direct a cold draught on to the pigs. See the two figures overleaf (bottom).

Fan Protection

Normally a cowl that protects the fan from a direct wind is sufficient. On some sites additional protection to stop upward eddies is required. It is worth putting in some form of back-draught shutters so that when the fan is off no air passes

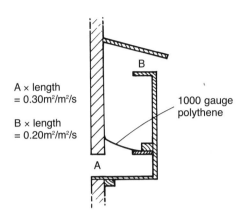

A × length
= 0.30m²/m²/s

B × length
= 0.20m²/m²/s

B

1000 gauge
polythene

A

Diagram showing a baffled outlet suitable for a pig/poultry house.

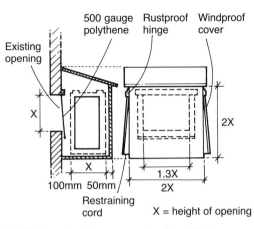

500 gauge
polythene

Rustproof
hinge

Windproof
cover

Existing
opening

X

2X

X

100mm 50mm

1.3X

2X

Restraining
cord

X = height of opening

Baffled outlets for pressurized ventilation suitable for a pig/poultry house.

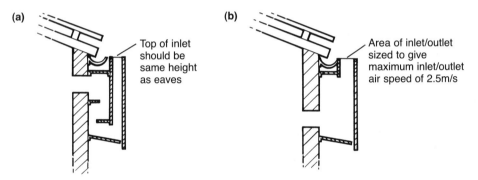

(a)

Top of inlet
should be
same height
as eaves

(b)

Area of inlet/outlet
sized to give
maximum inlet/outlet
air speed of 2.5m/s

Inlet and outlet designs, (a) light-proof type, and (b) non-light-proof type. (Farm Electric, 1990)

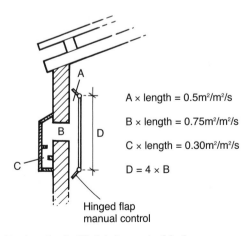

A

B

D

C

A × length = 0.5m²/m²/s

B × length = 0.75m²/m²/s

C × length = 0.30m²/m²/s

D = 4 × B

Hinged flap
manual control

Design for baffled inlets suitable for pig/poultry housing.

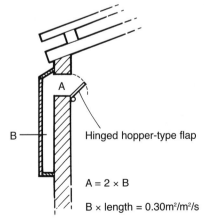

A

B

Hinged hopper-type flap

A = 2 × B

B × length = 0.30m²/m²/s

Baffled inlets suitable for a pig/poultry house.

through. Lightweight flaps of 20-gauge aluminium or plastic are sufficient, with a wire grid to prevent the flaps from over-turning. See the next figure (right).

THE REQUIREMENTS OF VENTILATION

As indicated earlier, ventilation may be based on the extraction at the roof and air introduced at the sides, forced in under pressure by the fans to find its way out at the sides or to be pulled across the house, from end to end or from roof to floor. It is probably correct to emphasize that it is not the actual system that is important but rather that it is properly designed, controlled and operated. A system should be in place that will function if the electricity system fails. The incoming air should reach the animals at low velocity in cold weather and at ambient air temperature when it moves over the animals; uniform velocity should be maintained at all times. With the most common ridge extraction systems, an inlet velocity of 60m/sec is achieved by allowing 0.5sq m of inlet area for each 1700cu.m/hr extracted (Sainsbury, 1992). The air inlet speed is possibly more important than the air inlet direction, but where the air entry is at the side of the house, the actual direction of it should be controllable. When the ambient air is cooler than that outside, the air ought to be directed upwards and away from the floor. If fans are blowing the air in, then baffle plates should be present to prevent it from causing draughts on the stock. Indeed, as long as these requirements are carefully met, inlet areas as low as 0.18sq m per 1700cu.m/hr are perfectly satisfactory.

A number of other requirements for a

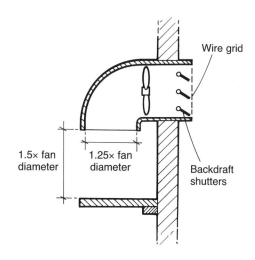

Detail for the fan intake for a pig/poultry house.

ventilation system may be outlined (based on Sainsbury, 1992):

- the air should enter/leave the house around the walls or along its length; there should be no dead spots created by large gaps between the ventilators
- the system should be able to cope in a semi-automatic way, with a very wide range between maximum and minimum ventilation rates, for instance, a ratio of 100:1 as in a broiler house, where a finished bird requires 7cu.m of fresh air an hour in summer; a day-old chick in winter needs no more than 0.08cu.m/hr
- it is usual to have a number of fans and speed controls and the fans in use may be progressively increased in number and speed from the minimum; it is important that the fans are speed-controlled or in some cases have controls that switch the fans on and off in series, this can give systems that are economical in cost, reliable, cheap to run and efficient

- air movement may be detected in the house by using smoke pellets or tubes, as mentioned earlier, these will indicate the air currents that are occurring inside the house.

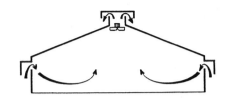

Ridge extract, side inlet ventilation system. (Farm Electric, 1990)

VENTILATION SYSTEMS

There are many different forms of ventilation and they are probably best described in the way that air enters and leaves the building.

Air Extraction Systems

Side Extraction and Ridge Inlet
This works by the fans situated in the sides of the building drawing air out through a ridge-mounted inlet. The system can suffer from the effects of wind interference blowing across the building. These systems can be better designed by exhausting the air up to eaves level rather than down to the floor. One advantage is that the fans are wall-mounted, which makes maintenance fairly straightforward. It is used for many pig (mature and growing stock) and poultry buildings.

Ridge Extraction and Side Inlets
This system consists of ridge-mounted fans with sidewall inlets. Sometimes the inlets are fitted with shutters, which allows for some manual adjustment. This technique of ventilation may be susceptible to wind effects (direction and strength). Good results have been obtained with an open-top baffle (with a rain gap at the base – see diagram above) whereby air is allowed to enter from gutter level as opposed to floor level as normally expected. The method of ventilation in general tends to be overused because of its simplicity.

Cross-flow Systems
Many pig buildings, having low, flat roofs, use this method. There is an airflow across the building, between the outlets and the inlets, at opposing sides of the

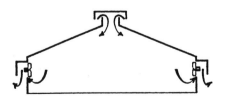

Side extraction, ridge inlet. (Farm Electric, 1990)

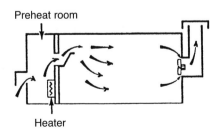

Crossflow ventilation. (Farm Electric, 1990)

building. Caution is needed to avoid cold draughts from the incoming air, often achieved by preheating the incoming air. The system may be susceptible to winds which is a drawback. It is used in package pig-weaner buildings (for instance, flat decks) and in small room conversions.

High-speed Jet System
In essence, this is similar to the ridge extraction/side inlet system, but it has the advantage that the internal air flows are controlled to suit different ventilation rates. In fact, the system may have either ridge or eave inlets, with variable apertures to suit the volume of air going through the fans. For example, as the controller decreases the ventilation rate the inlet gap is narrowed to maintain a constant air speed (5m/sec [16ft/sec]). The design of the system allows the incoming air to cling closely to the inner surface of the building; this will reduce cold draughts. The system manages to maintain well-defined air patterns no matter what the outside conditions are, which is

an advantage. Therefore this system of ventilation is certainly more wind-proof than the other methods mentioned. This method of ventilation is used quite successfully on large, widespan pig and poultry buildings up to 24m (79ft), provided that the layout is open, without solid pen divisions and that the inside surface of the roof is relatively smooth.

Air Pressurization Systems

Ridge Pressurization and Wall Outlets
In this system the fans are mounted in the ridge and blow air (from outside) throughout the building, eventually exhausting it via the outlets which are situated in the walls. This is fairly popular since, with roof pitches of less than 15 degrees, it is not susceptible to the wind. If the wind blows against one side of the building as long as the outlets have back draught shutters, those on the windward side will close and those on the leeward open, providing an exhaust for the air. One of the main problems of a pressurized system is the distribution of incoming air from the fans without causing draughts. The systems may be advantageous in older, 'leaky' buildings when gaps can act as inlets which cause draughts when the fan acts as the extractor, that is, sucking.

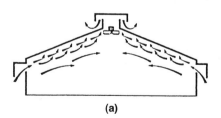

(a)

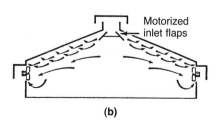

Motorized
inlet flaps

(b)

High-speed jet ventilation, (a) eaves inlet and (b) ridge inlet. (Farm Electric, 1990)

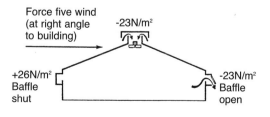

Force five wind
(at right angle
to building)   -23N/m²

+26N/m²   -23N/m²
Baffle   Baffle
shut   open

Ridge pressurization, wall outlet. (Farm Electric, 1990)

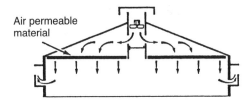

Pressurized ceiling. (Farm Electric, 1990)

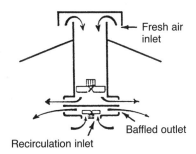

Baffled fan with air mixing. (Farm Electric, 1990)

Pressurized Ceiling

In this system an air-permeable material such as glass fibre forms a false ceiling. The function of this is to spread and reduce the speed of the incoming air. The other advantage of using a material such as glass fibre is that it acts as an insulator against heat loss. With a thickness of usually 60mm (2.4in), the maximum air speed through it is no more than 0.15m/sec (0.5ft/sec). When designing the system, therefore, as a guide 1.85sq m (20sq.ft) of ceiling area is required for every 1,000cu.m/hr (1,300cu.yd/hr) of ventilating capacity. It is also necessary to seal all other surfaces with polyethylene film to prevent unwanted leakage. One major disadvantage is that the ceiling tends to clog with dust, which ultimately reduces the air flow. Hence the panels must be easily removable for cleaning purposes. The system is used for many purposes, particularly for young stock or where an even temperature is required.

Baffled Fan with Air Mixing

This system incorporates a pressurizing fan which blows down a short tube on to a horizontal air deflector and out sideways through baffled guide vanes. A recirculation fan is located on the underside of the deflector and draws up internal air from the building on to the lower side of the air deflector and, again, exhausts sideways.

Both these sets of air mix as they leave the unit. A major benefit of the system is that in winter, where cold air would drop directly down from the fan inlet causing a cold draught, this is prevented. The pressurizing fan is speed-controlled in order to deliver enough air to maintain the room temperature. The recirculating fan can either be left to run at full speed all the time or speed-controlled where speed is increased as fresh air delivery is reduced, so that the slow, fresh air is mixed properly. The main application for this system is in new buildings and building conversions for poultry and pigs. One situation where it would not be recommended is for very small rooms or buildings with solid pen divisions.

Recirculation / Ventilation

With these systems, instead of controlling the amount of incoming air by varying the fan speed or the number of fans, separate fans running at higher speeds are used. The amount of incoming air is controlled by a damper which can change the proportion of recirculated to fresh air. This technique can either be a pressurizing or a depressurizing system, often supplied as a factory-assembled unit. The overall performance tends to be good because the fans are working at higher speeds when the ventilation rate is low (unlike other systems described here). Also, since cold air is mixed with the recirculated internal air, draughts are fewer. See the accompanying figures. Some systems are designed to exhaust and draw in fresh air at the same point. One advantage of these systems is that they are inherently windproof since there is no wind pressure differential between the inlet and the outlet. A relatively cheap way to distribute air from a pressurizing fan (or recirculating) system uses a slotted polyethylene duct.

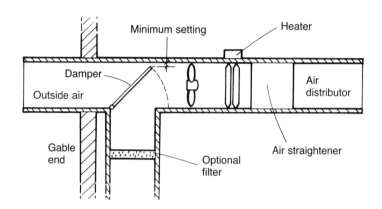

Typical recirculation unit. (Farm Electric, 1990)

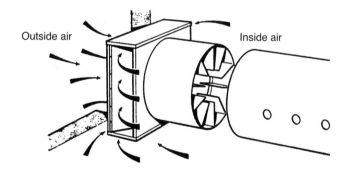

Balanced flue type recirculation system (Farmex). (Farm Electric, 1990)

Airflow with ACRU unit. (Farm Electric, 1990)

Controlled mixing Extraction Complete recirculation

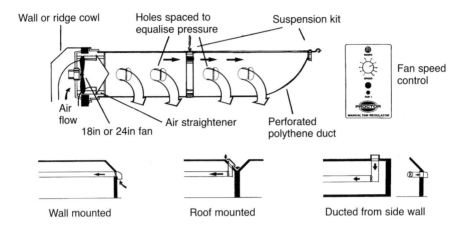

Typical ventilation system as suitable for calves, for example, using a large ridge-mounted, polythene tube with a fan. (Proctor, 1997)

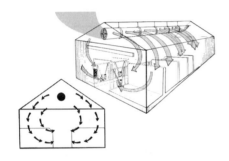

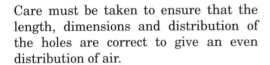

Two diagrams of the same system as shown in the preceding figure, showing how air movement is created. (Proctor, 1997)

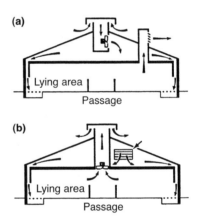

Cross-section of piggery with down-jet ventilation system, (a) pressurized, and (b) exhausted. (Farm Electric, 1990)

Care must be taken to ensure that the length, dimensions and distribution of the holes are correct to give an even distribution of air.

Down-jet System

This involves the introduction of air into a building through long, narrow slots or down-jets at the edges of the room. The slots are 50mm (2in) wide to allow a maximum air speed of 4–5m/sec (13–16ft/sec), and the system may either be pressurized or exhausted down-jet. These systems offer a simple way to give, under any external weather conditions, a defined, lower-temperature dunging area nearer the sidewalls. The down-jet does not require inlet modulating devices, as used in the high-speed jet system. The main use for this system is for partially slatted pig buildings where the slats are along the walls of the building.

(These descriptions are based, with permission, on *Controlled Environments for Livestock* (Farm Energy Centre, 1990).

CHAPTER 12

Methods and Techniques for Measuring the Environment

Once the livestock building has been designed with inlets, outlets and possibly fans to control the atmosphere, it is vital that checks may be performed to check the environment. These measurements may be made with relatively simple equipment, and then there should always be the opportunity to make minor adjustments to the environment. The following descriptions are of equipment that is readily available for any farmer who wants to gauge the performance of his building.

TEMPERATURE AND HUMIDITY

Probably the easiest form of instrument used here is the sling psychrometer. This resembles an old-fashioned football fan's rattle and contains two thermometers, one of which has a wetted gauze around its bulb. These are used to measure the wet and the dry bulb temperature. When using the instrument, it is important to rotate it at a steady speed and then take a reading of the two temperatures as quickly as possible.

The dry bulb and the wet bulb temperature can be used to find the humidity level by referring to a chart or slide rule, usually supplied with the instrument. The outside temperature and the humidity should also be measured with this instrument; if the outside temperature is fairly high and the air is very humid, then the conditions in the livestock house may not be very different. Once a series of readings at several points in the building have been taken a plot of temperature and humidity can be made. It is important to remember that readings must be taken at stock level since it here that measurements are important. Cold spots and high humidity areas within the building can be pinpointed and then corrective action may be taken.

It is important that the instrument is kept clean and it should be protected in its original case when not being used. The two plates overleaf show the instrument in use.

As an example, if the wet bulb temperature is 18°C and the dry bulb is 22°C, then the relative humidity is 68 per cent (the psychrometric chart may also be used to determine the humidity from the two temperatures). Note that if the atmosphere is completely damp, as in a bathroom following a shower, then the relative humidity (RH) will be 100 per cent. The precise control of humidity may not be that important, but extremes must be avoided, for example, in pig housing we are usually looking for 50–70 per cent RH. If the atmosphere is too dry or too wet, then health problems can occur.

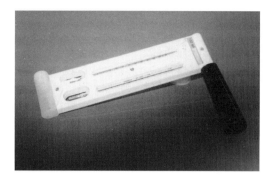

ABOVE: *A whirling hygrometer: wet and dry bulb thermometers are moved around in the air to obtain a more accurate reading of relative humidity.*

RIGHT: *A picture showing the whirling hygrometer in use.*

Observations are also an important means of checking on the level of dampness or otherwise, thus mould and condensation on walls indicate a damp problem.

High humidity may be caused by the following conditions:

- under ventilation in winter (insufficient air movement and air exchange)
- post-cleaning, as in all-in and all-out pig systems
- damp straw
- new concrete not fully cured
- leaking drinkers/water lines
- poorly designed floors and drainage systems
- misuse of spray cooling systems
- unvented gas heaters (which contain high levels of moisture).

AIR FLOW

Smoke can be used to make the air visible and so to detect flow. It may be produced in small amounts by using a smoke tube or in larger amounts by smoke pellets. Alternatively, an air-speed measuring instrument, such as a hot-wire anemometer or a digital anemometer, may be used.

A smoke tube comprises a glass tube which contains a light powder which can be expelled by using a rubber bulb once the tube ends have been broken off. Several presses of the bulb discharge a thin stream of smoke which then remains visible for 2 or 3sec. The smoke can be used to indicate the direction of the air flow and will give the operator an idea of its speed. It can be used to advantage around fans, doorways, inlets and outlets and other points where air flows may occur. Smoke pellets, originally designed for boiler and chimney observations, give copious amounts of smoke and show the total movement of the air in a building. Large amounts of smoke can show the efficiency of air mixing within the building and indicate where draughts can occur. Smoke can also show where still air pockets are and demonstrate the effects of changing the pen divisions, for example.

But it must be borne in mind that smoke pellets are unsuitable for use when livestock are in the building.

Anemometers are more expensive instruments and may not be justifiable in all types of livestock housing. The hot-wire anemometer may be used to check air speeds from 0–30m/sec (98ft/sec) and air temperatures from 0–40°C, and the digital instrument is normally used to check air speeds of typically 0.4–35m/sec (1.3–115ft/sec), for example, the correct functioning of fans. It is difficult to give exact air speeds for different situations, but, with young animals that are more susceptible to draughts, a maximum air speed that should not be exceeded is 0.25m/sec (1ft/sec). The plates below and overleaf show the use of anemometers to measure air flow.

RIGHT: The air flow tester and smoke pellets.

BELOW LEFT: The air flow tester (Drager). The glass tubes contain an inert, porous material impregnated with fuming sulphuric acid. When air is forced through the tube, smoke is produced, this is carried by the air currents showing ventilation air movement.

BELOW RIGHT: Smoke pellets: the pellet is rested on a suitable surface and lit with a match or lighter. The resulting air flow (which lasts for about 1min) will show air ventilation movement.

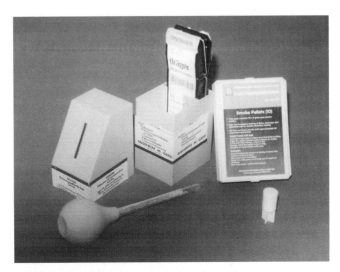

LEFT: *The digital anemometer and the hot-wire anemometer as used to detect air movement.*

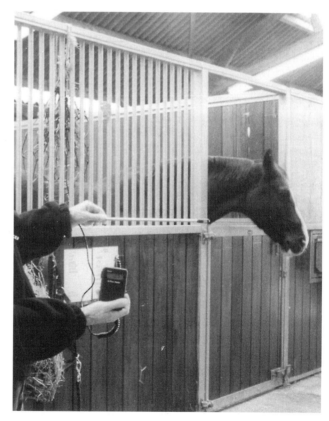

The digital anemometer. This instrument will measure air speed from 0.4 to 35m / sec, and so is ideal for detecting air movement.

The hot-wire anemometer used to ascertain air speed (and temperature), a very accurate instrument that will measure down to 0.1m / sec and air velocity up to 30m / sec.

Excessive air movement may occur due to poor ventilation design, wear and tear on equipment and draughts caused by leaks through or around doors and windows (young animals are particularly susceptible). Inadequate air movement may occur through lack of attention to fans and their maintenance (where applicable), blocked inlets/outlets, bad positioning of other equipment; for example, fluorescent lights, external wind patterns (these can seriously upset ventilation systems) and the position of other buildings nearby.

UNWANTED GASES

The atmosphere in a livestock building may be tested by using a colour indicator which can give a rough estimate of the percentage of the gas present. The indicator, which is contained in a glass tube, changes colour as air from the building is drawn over it by squeezing a small bellows. The extent of the colour change indicates the amount of gas present, and a different tube is used for each gas being tested. Many different tubes are available, but it must be remembered that these can be used only once, and they are not particularly cheap.

Typically, carbon dioxide tubes (0.01 per cent, 10 pumping strokes) can be used to check the atmospheric concentration in a livestock building; the outside air should be around 0.03 per cent and inside it may be slightly higher, depending on the housing conditions. Of course, if the inside level goes unacceptably high then breathing becomes more difficult and a very high concentration indicates a lack of oxygen.

Other, small, hand-held instruments are available for measuring carbon dioxide levels; there is one that includes a data-logging facility which can store up to 800 readings.

CONTINUOUS RECORDING

It is rarely sufficient just to measure the conditions inside a house once and at only one point. This can give a false reading which may vary hugely over the time of day, or night, the stocking density and the type of building, for example. For this purpose, data loggers of one sort or another can be useful. They are probably not the sort of instrument a farmer would own himself but one can be most beneficial to the consultant who is trying to understand what is happening inside a building over, say, 24–48hr or perhaps up to a week. Modern data loggers are available as very small instruments and are in plastic tubes, around the same size as camera film cases. Often they are sold as single-channel data loggers, for instance, a Tiny Talk. These can be programmed by computer to log the temperature, for example, at 30min intervals throughout the period, placed discreetly where the logging is to take place and then be downloaded on to a computer to produce the statistics required.

Another type of instrument is the Squirrel Digital Meter/Logger, which consists of a box with outlying leads plus sensors. This is more expensive but it will log several conditions in addition to temperature. It can record temperatures between -30 and 65°, with humidities up to 95 per cent. There are several channels which the machine will record from and so temperatures could be recorded from several positions inside a building and on the outside as well. The plates overleaf show instruments that can be used to measure aspects of the environment.

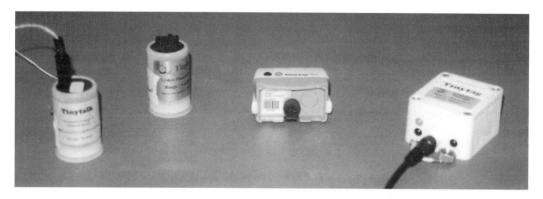

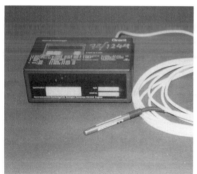

ABOVE: *A variety of 'Tiny Talk' dataloggers that may be used for environmental assessment. These will record aspects of the environment such as temperature over a certain time period such as a week and then download on to a computer to give a graphical output of temperature vs. time. They are essential tools for the building specialist.*

LEFT: *An alternative datalogger that may be used to detect trends in temperature (and other environmental data) over a period such as a week. This may be the only method of showing whether a particular pen at the corner of a building is subject to draught, for instance.*

Care must be taken with the data loggers themselves – animals will damage them very easily, and dust settlement may be a potential problem in some houses. Moisture, heat sources or simply theft must also be avoided where possible. Direct sunlight falling on a sensor must be avoided, as should be any position where a light bulb could shine directly on one. Although the type that is located inside a plastic case is free of contaminants, the one that relies on sensors and leads may be less robust.

PSYCHROMETRIC CHARTS

The psychrometric chart (see the figure on p.107) shows the relationship between the following variables:

- dry bulb temperature; this is the ordinary air temperature, and is read from the horizontal scale on the base line
- wet bulb temperature; this is read from the sloping straight lines running from the saturation line (that is, 100 per cent saturation); it is so called because the bulb of the thermometer is wet when the reading is taken, it is accomplished by slipping a cloth 'sock' on the bulb end and then dipping it in water, the moisture evaporates from the sock and consequently cools the surface of the bulb; if the moisture content of the air is low, evaporation from the sock takes place rapidly and the wet bulb temperature reading is lower; air with a low moisture content therefore has a lower wet bulb, and air with higher moisture content has a high wet bulb in relation to its dry bulb; when the

relative humidity is 100 per cent, the wet bulb temperature is the same as dry bulb value
- dew point temperature; the temperature at which moisture condenses on a surface
- specific volume (cu.m/kg); the volume occupied by a given weight of air depends upon the temperature of the air and the barometric pressure at standard sea level pressure; the volume of 1kg of dry air at a given pressure will vary with the temperature
- specific enthalpy (kJ/kg); air has both sensible and latent heat and the total heat content of the air at any state is the sum of the sensible and the latent heat; sensible heat is a function of dry bulb temperature; the amount of latent heat in any given quantity of air depends upon the weight of vapour in the air and the latent heat of vaporization of water corresponding to the saturation temperature of water vapour
- percentage saturation or relative humidity; this is read from the series of curves running from top right to bottom left and is the actual amount of moisture in the air compared to the maximum moisture the air can hold
- moisture content (kg/kg); this is the mixing ratio and is read from the right-hand vertical scale
- vapour pressure (measured in millibars); this is read from one of the right-hand scales, where it is present.

Note that the saturation curve represents 100 per cent relative humidity and that at this condition, the wet bulb, dry bulb and dew point temperatures all have the same value. Probably the best way to understand the chart properly is to have a look at some examples.

1. Take 5kg of air at a dry bulb temperature of 30°C and a wet bulb of 22°C.
   (a) What is the relative humidity (RH)? What is the weight of water per kg of dry air? What is the specific volume?
   (b) If the air is warmed to a dry bulb temperature of 40°C, with no addition of moisture, what is the new RH and specific volume?
   (c) If the air is now cooled to have a RH of 100 per cent (dew point), what is the dry bulb temperature?

   (*Answers:* 48 per cent, 0.0132kg/kg, 0.877cu.m/kg, 0.906cu.m/kg, 18.5°C)

2. 1kg of air is at 31°C and a RH of 50 per cent.
   (a) What is the wet bulb temperature, the moisture content and specific enthalpy?
   (b) If the enthalpy is reduced by 10kJ/kg (with no change in water content), what are the new conditions?
   (c) How much more heat would need to be removed for the air to reach its dew point?

   (*Answers:* 23°C, 0.0144kg/kg, 68kJ/kg, dry bulb 27.5°C, moisture content 0.0118kg/kg, specific volume 0.867cu.m/kg, 68 – 47 = 21kJ/kg)

3. 10kg of air at 25°C and 90 per cent RH is warmed to a RH of 50 per cent (with no additional water). What is the quantity of heat required?

   (*Answer:* 74 – 68 = 6kJ/kg)

CHAPTER 13

Types of Fan and Air Flow

FANS

There are several different fan varieties and all come in several sizes. The speed of operation may be very slow to very fast, and their performance will also vary tremendously. In definition, 'a fan is a rotary bladed machine which delivers a continuous flow of air or gas at some pressure, without materially changing its density' (Cory, 1991). A fan has blades of some kind attached to an impeller and these exert force on the air, maintaining the flow through it and raising the pressure. A fan characteristic curve may be drawn for any fan, showing the pressure (displayed vertically) vs. flowrate (along the baseline).

A fan's efficiency will be at a maximum at some point along this line, and the power consumption for a given duty is then at a minimum. At this point the noise level is also usually low. The fan may easily be operated at other points along the line, of course, but:

- its efficiency will be lower
- the noise will be higher, or
- the purchase price of the fan will also be higher.

Some parts of the curve can also lead to situations such as motor overloading, inadequate cooling, unstable operation and/or excessive vibration.

Types of Fan

There are three main types: axial flow, radial or centrifugal and the propeller fan.

Axial Flow Fan

This is well established for air conditioning and many other applications in agriculture. The capacities range from 0.5–150cu.m/sec (0.65–196cu.yd/sec), with pressures up to 1.5kPa. Higher pressures can be obtained by multi-staging, that is, having fans in series, with higher volumes being obtained by having units in parallel with each other. The blades of the fan are usually cast in an aluminium alloy in order to make them light and they are of the non-overloading type. For building ventilation, a single-stage type would normally be adequate.

With multi-staging, contra-rotating fans are most suitable for series connection. Each impeller is driven by its own motor and they rotate alternately in opposite directions. Therefore each impeller effectively cancels out the swirl of the one before, and guide vanes are unnecessary. The fan pressure of a contra-rotating pair is 2.5 times that of a single stage. The blades of alternate fans are different from each other, that is, one has a left-handed screw and the other a right-handed one. Note that the reversal of the axial fan direction reverses the air flow. But the performance of the guide vanes in reverse is relatively poor, and a reversed fan may produce only 60–70 per cent of the forward volume.

The blades are normally directly coupled to the shaft of the electrical motor, making the unit more compact. The air leaving a single-stage fan is a spiralling column rotating in the same direction as the impeller. The fans are well suited for use inside a ducting but can be very noisy, giving a high-pitched whine. Hence they may not be applicable for some types of livestock housing. In broad terms, they have a high efficiency with a low power requirement.

The output of an axial flow fan can be improved in the region of 5 to10 per cent by using a bell-mouth inlet to smooth the passage of air into the fan itself. This is essential in order to achieve the best performance. These fans are normally used in crop drying and would not usually be adopted for building ventilation purposes.

Radial or Centrifugal Fan
These have a number of radial blades on the periphery of an impeller. The air held between the blades is thrown outwards under centrifugal action and replaced by air entering from the centre of the casing. Further rotation results in this flow of air being continuous and a sustained pressure being set up. The fans give a medium to high pressure development and are usually belt-driven for flexibility in speed selection.

Multi-staging may be carried out with these fans, usually for low-cost applications, giving limited tip speed. A variation on this type of fan is the cross-flow which has an impeller similar to the centrifugal type. One example of this may be found in domestic vacuum cleaners for economy or wide air curtains above shop doors. These fans are less efficient than the axial flow types and need to be physically larger

than them in order to move the same amount of air. They do, however, produce lower noise levels. They are suited to special industrial uses such as grain driers, pneumatic conveying and hay drying. They are not normally used for ventilation systems.

Propeller Fan
These are the simplest of fans but are capable only of low pressure development, for instance, up to 100kPa, with 0.3–10cu.m/sec (0.4–13cu.yd/sec) flow. A common description is that the impeller resembles a ship's propeller. The blades on these fans are usually non-adjustable, and the fans may have three, four or more wings; probably the most common type has the four-wing blades. The concave side of the blades is the output or upstream side. The impellers are mounted on the shaft of the driving motor. Air flow effectively cools the motor, which may be totally enclosed in order to exclude any dust. The fan needs to be protected from wind interference and any obstructions in the duct work must be avoided as these reduce the volume of air handled by the fan. Within certain margins, the faster the fan rotates, the greater the volume it delivers and the greater the fan noise. Also, the pressure of air is proportional to the square of the fan speed, and the air speed reduces in proportion to the cube of the distance from the outlet. This fan is the most suited for livestock ventilation uses, with relatively low costs and low noise levels. One big advantage of the propeller fan is its cheapness. It is suitable where large volumes of air have to be moved against a relatively small resistance to air flow. Most livestock ventilation systems use single phase fans of up to 630mm in diameter, but three-phase fans are available for larger installations.

AIR FLOW AROUND BUILDINGS

If the wind is blowing square on the building, with the effect that it is deflected around the ends and roof, it creates a suction on these areas. The faster the wind speed, then the more will be the suction. The sides of the building will also experience suction effects, and these will be greatest near the windward edge. Any doors or windows in these walls will be subject to the same effects and damage to these can easily occur. As the wind blows between two buildings it can cause a similar suction effect on the sides facing the gap.

As far as the roof pitch is concerned, on the windward slope the pressure depends on the pitch. If the roof angle is below around 30 degrees the windward slope can be subjected to suction; if the roof is steeper than 35 degrees then there is generally a positive pressure developed. Leeward slopes are always going to be subject to suction; if the wind blows along the direction of the ridge, then severe suction can occur. Indeed, the forces due to suction on a low pitched roof, like the type frequently used in agriculture, may be most severe. The uplift on the roof may be far greater than the roof self-weight,

which may result in the roof sheets' being lifted off at their corners in high winds.

There is no uniformity of suction or pressure over a wall or roof surface; pressure is generally greatest near the middle of an area and decreases near the upwind edges. The most pronounced suction occurs at the corners and along the edges of roofs and walls. This is often a problem in agriculture where careful attention must be paid to fixings at these locations.

If there is a projecting feature, such as a chimney stack or dormer window, this may create eddies in the airflow which then cause local loads on this feature, as well as having some modification on the wind loads on the roof in the vicinity. The roof cladding around such projections requires special attention to cope with potential problems. It is also the case that roof overhangs will be subject to an upward pressure.

If the wind blows obliquely (not at a right angle) it is deflected round and over the building. The pressures on the walls are generally less strong compared with the situation where it blows square-on, but vortices develop. Again, these result in high suction at the edges of the roof, which must be resisted by the appropriate fixings. Most reports of roof damage are, in fact, caused by this effect.

CHAPTER 14

The Conversion of Existing Farm Buildings

During the slump a number of farmers put up their farms for sale, and found it impossible to dispose of them, at least at a price that would cover the inevitable mortgage. Then a firm of house agents in a neighbouring town conceived the idea of buying these farms and dividing them up into lots. The farmhouse, usually old and picturesque, would be sold by itself with its orchard and garden and a small shed or two. The oasts and barns would be sold to enterprising people with a passion for the quaint, for conversion at vast expense into weekend retreats.

S. Kaye-Smith (1937), 'Laughter in the South-East', in Clough Williams-Ellis (ed.), *Britain and the Beast* (Watkins and Winter, 1988)

CHANGE OF USE

'Change of use from an agricultural to non-agricultural business will generally require planning permission' (NFU, 2001). Most local authorities will generally be supportive of a concept of a change of use; there will, however, be some problems to overcome, such as environmental or highway issues. It may be a case of studying the council's local plan together with a discussion with the planning officer before any alternative business is contemplated. The change of use will normally commence a charge of 'business rates', and, while it is possible to pass this charge on to a tenant, there may be a future unwanted liability if the building eventually becomes vacant.

In many situations a change of use is what is required to keep the building in everyday occupation; indeed, from a legal point of view, it is generally considered that the best way to retain a farm building is to keep it in use. As buildings become disused and redundant they ought to be converted to some other purpose in order to keep them in active use. There are many examples around the countryside of farms being converted to alternative uses; this is happening almost nationwide as farmers look to make a 'profit' on something that is outside normal farming practice. Their incomes have fallen away to what most people would think is a danger level, and so they are on the lookout for other enterprises that might bring in some extra cash.

In order to make a judgement on what is most suited for a particular building, one has to take into consideration all the local factors, in conjunction with the thoughts and desires of the local authority. Of course, there will be the difficult task of balancing the desires of the landowner with the local, wider sense. It is necessary to identify the most applicable use of the building that would be compatible with its fabric and setting; this may not always be the most profitable use. Sometimes an alternative use for a farm building will require 'less damage' to be done to it than some other, more modern agricultural use.

Forty per cent of agricultural holdings in England and Wales are believed to have used land, buildings, water and personal skills to pursue new ventures outside farming. Some of these activities are conventional, such as farm tourism and dairy product manufacture. Others are new and even novel services, unrelated to the farm (ADAS, 1991). It can be assumed

to have changed considerably since 1991, and now the figure of 40 per cent must be more like 60.

There should be no reason for preventing the 'conversion of rural buildings for business re-use, provided that:

(a) they are of permanent and substantial construction
(b) conversion does not lead to dispersal activity on such a scale as to prejudice town and village vitality
(c) their form, bulk and general design are in keeping with their surroundings
(d) imposing reasonable conditions on a planning permission overcomes any

Picture showing a traditional farm building, now unused and falling into disrepair. These buildings are unsuitable for modern farm machinery but could have alternative uses for different livestock or diversification projects.

Another view of the same building. Notice the roof slates falling off, allowing water to penetrate and leading to decay and timber rot.

A farm building now unsuitable for large modern machinery. It is too low with doorways too narrow.

A farm building undergoing alterations to convert it from traditional uses to an alternative function not directly connected to agriculture.

legitimate planning objection (for example, on environmental or traffic grounds) which would otherwise outweigh the advantages of re-use

(e) if the buildings are in the open countryside, they are capable of conversion without major or complete reconstruction.' (Department of the Environment, PPG7, 1997)

It can be argued that most problems associated with the conversion of rural buildings are centred on three main issues: the redundancy of the stated building, the character and location of the building and the quality and design of the proposed conversion.

Redundancy is always the main concern, and a rural building must be shown to be redundant if it is to be reused for an alternative purpose. This can give rise to certain problems. The main reasons for redundancy of farm buildings may be given as being that it is unsuitable for modern farm practice, in the wrong location for agricultural use or it is severed from the farming unit (*Farmers' Weekly*, 10 Jan. 1992).

CONSERVATION OF FARM BUILDING REPAIRS

In order for a farm building to be converted, there may be considerable work that must be done. For example:

- the timber-frame walls will need insulation
- the roof will require insulation
- solid walls of brick or stone will need a damp-proof course to prevent rising damp
- walls may need an inner lining
- fire protection, for example, to roof trusses
- services such as gas, oil, water, electricity, telephone and computer cabling
- drainage for rainwater and sewage (to soakaway and septic tank).

There have been suggested over the years a number of principles for conservation. These may be listed as:

- a thorough understanding of the history and characteristics of the building as a necessary preliminary to its repair
- with the aid of this understanding, the causes of defects needing repair can be established and a plan for remedying them prepared
- any repair should be the minimum required to conserve the building and ensure a sound structural condition for long-term survival
- the repair is intended to restrain the process of decay, but not at the expense of the character of the building nor involving any unnecessary disturbance or destruction of the historic fabric
- unless the existing fabric has failed because of inherent defects in design, choice of materials or opportunity for maintenance, the selected repair techniques should be compatible with existing materials and methods of construction
- honest repair should avoid artificial ageing but should not be unnecessarily obtrusive. (Brunskill, 1999)

DIVERSIFICATION – THE IDEAS

Before any diversification scheme is started, there are some rules which must be considered. These are:

- you must ensure that your current farming business is run well; it will probably be worthwhile getting independent financial advice to inform you whether or not you are making a reasonable profit. If the existing enterprise is making a loss, then is it really feasible to diversify?
- it may be possible to exploit under-used resources that could be used advantageously with the minimum of risk
- it is necessary to clarify your objectives; it will be pertinent to ask the question: what is your personal ambition and what do you desire to get out of the 'new' business?

A feasibility study will be necessary to determine whether the enterprise chosen will be financially viable; the main features of this investigation will be:

- project description
- market research and strategy – ascertaining the customer's needs.
- capital investment
- project income and cash flows
- labour
- effect on existing business; many farmers forget that much of their time will be taken up in running the enterprise

- statutory/planning requirements (the planning requirements for a diversification scheme may be quite complex and involved and certainly early advice must be sought from the local planning authority, LPA)
- conclusions and recommendations; firm conclusions must be drawn as to whether to proceed. (ADAS, 1991)

The following examples may help to give some idea as to what many farmers are now considering (and have been doing for several years) for diversifying purposes:

- processing of farm produce and craft manufacture; this category might include the production of pottery, weaving, textiles, treatment or bottling of spring water, or the repair and renovation of old farm machinery
- rare breeds; many farmers have taken the route of diversification into one

RIGHT: *Converted farm buildings, now in use as a visitor centre.*

BELOW: *A recently converted farm building with a thatched roof.*

aspect of keeping rare breeds, this might be pigs for example, where they may be kept and reared in buildings that are considered to be totally inadequate for modern production

- farm shop, the produce for which has primarily been produced on the farm
- direct 'pick-your-own' enterprises; the feasibility of these projects will depend on the proximity of roads and paths, also fences, hard standings and carparks will be necessary
- catering; many types are considered for diversification, for example, from the basic tea bar to restaurants and canteens; an issue to be taken into account here is whether a licence is required for alcoholic beverages, if this is so, then this is another hurdle to overcome; much equipment will be necessary, including freezers, cold storage cabinets and cookers
- accommodation; most of this when completed has to comply with the English Tourist Board (or equivalent) specifications; certain minimum standards are laid down with regard to all fixtures and fittings while different standards are set for camping barns and bunkhouse barns and camping and caravan sites
- sports, recreation and education facilities; sports may include football, table tennis, water sports and swimming but also included are field sports such as clay-pigeon shooting, 'war games' and crossbows; recreation includes play areas, nature trails, picnic areas and ponds for angling; education may involve farm museums, craft workshops or nature trails; there will be many facilities to provide under this category of diversification
- livery and horses/ponies for hire; livery may be taken as the provision of accommodation and the care of other people's

horses on a contractor basis; this will include the provision of stabling and other buildings, together with an electricity supply
- letting of eligible facilities; this method provides the owner with an income so that he can let his land or buildings to another person.

LOCATION OF THE DIVERSIFICATION ENTERPRISE

The location is all-important, and a farm shop that is visible from the road with easy access and parking and located in an attractive and well-lit building will draw potential customers. It will do much better than one that is out of sight, around the back of a muddy yard and in a dingy shed. The type and the shape of the land have a big bearing on the enterprise chosen; someone wishing to set up a shooting school would be better off if there were an old chalk pit surrounded by scrub or trees than if he had chosen bare, arable land. Another example might be a farm on the edge of an urban area which could seem like the perfect place for a pick-your-own enterprise. However, the soil type has an even stronger bearing – it might be totally unsuitable for the crops being grown and be sticky and totally unattractive to visitors.

If the buildings were located on the edge of a town and had an attractive appearance with outline planning consent, they might be actively sought by agents and developers alike. Perhaps less desirable farm buildings located outside a village might find buyers that were seeking premises for light/services industry. Indeed, a country location can offer other advantages: a pollution-free

environment, low purchase cost in relation to city buildings, peace and quiet for
offices being the most important.

Farms are often situated well off the
road, and even the track leading to the
property could be very long, and obviously
need to be correctly surfaced. The construction of a major by-pass road might
alter access to a farm or perhaps spoil its
rural character; this would mean that the
farm would be less suited for other uses
that depended on peace and quiet such as
an art studio. Many farms have expanded
over recent years, usually by taking on
land that used to belong to another holding – sometimes a good way away. It is not
unusual therefore to find tractors and
combines driven for several miles to get to
this additional land. This can lead to
problems from the planning and highways authorities. Visitors might be
unwilling to find their way down winding,
narrow and often muddy lanes to get to
the attraction. There may be other features that need to be installed on the
farm, such as ramps for disabled access
and fire-prevention measures. It is worth
mentioning that if the farm were situated
in a National Park or Green Belt, for
example, it would be more restricted as to
what other developments could be carried
out when compared with a farm in a less
protected area.

The possible threat of intruders and
burglars becomes greater as more people
visit the farm following the creation of the
new enterprise. It becomes increasingly
difficult to control who actually comes on
to the premises, and the threat of the
theft of office equipment, cash and goods
in a shop or warehouse becomes very real.
Security measures will help, and in time
the farmer will get to know who may be
looking for trouble; certainly a visit from
crime prevention officers will go a long
way to pacify any unease.

BUILDING ALTERATIONS

If the buildings in question are 'attractive', then public access (for instance, to
craft shops, interpretation centres, tea
shops and self-catering accommodation)
may be applicable, whereas less attractive buildings could become a machinery
repair place, a rare breeds centre, workshops, some food processing and some
sports uses. If a building already has a
farm use but is particularly attractive it
may be worthwhile erecting a new building in order to free the old one for even-
tual sale or letting. As far as the work
required on the building is concerned, 'the
safest policy is usually simply to provide
a shell with a concrete floor, adequate
natural light, doorways (perhaps not even
doors until tenants' needs are established), water, electricity and washing/
toilet facilities, then leave the internal
fitting for consultation with tenants'
(Haines and Davies, 1987).

Roof Structure

The roof may well include trusses, rafters
and a wall plate. The renewal or replacement of the roof can lead to additional
loading, so care must be exercised if this
is the case. A good and sound cladding
without underlay is acceptable and
means that ventilation will not be necessary in a cold roof construction. Most old
buildings are reclad in old or new material to protect against driving snow or rain,
especially with tiles in exposed situations, or reclad because of the necessity
for other roof work. It is worth pointing
out that building control departments
and conservationists usually prefer warm
roof construction with an underlay and
insulation decking underneath, leaving
the required minimum 50mm (2in)

ventilation gap. This practice makes it possible to have exposed roof trusses and a sloping ceiling, thereby adding more space inside and more character to the conversion.

Walls: Foundations

Building control will insist on exposure of the foundations at various points around the building by using trial holes. This information is required by the authority, quite understandably, at an early stage. For old farm buildings the existing foundations are usually very shallow and considered to be inadequate by today's standards and the cost of underpinning will add considerably to the overall works bill. (As an aside, the foundations under many old houses would now be considered as being totally wrong and inappropriate. If someone dug down to lay foundations and found the subsoil to be the same at 300mm (12in) depth as at 2,000mm (79in), why dig deeper than 300mm? Surely this represents a waste of time and effort? It is certainly an interesting thought given that old houses have stood, most perfectly safely, for sometimes hundreds of years.)

It is sometimes possible to reduce the load and save some of the fabric, especially if the whole wall is inadequate. This can be done by supporting the roof with a steel frame, usually internal and effectively hidden or camouflaged; there are many examples where this has been done successfully. Sometimes there are proposals to deepen the floor levels, but these are rarely feasible because of shallow foundations. Creating sufficient headroom, say 2m (6ft) below the underside of roof trusses. for example, may be done by using a raised ceiling. It is normally possible to lift the whole roof without spoiling the character of the building. It is often a

good idea to avoid low ceilings since these can give rise to ventilation problems associated with modern uses for buildings. (Note that in many old buildings the ceilings are incredibly low, this is because people were shorter in stature years ago, and also the modern uses of buildings have drastically changed the requirements of the fabric, this will be noticeable when visiting old buildings.)

Walls: Fabric

Often the original structure of a building may be damaged irretrievably simply because its owners cannot find any way of keeping the walls as they are and retaining the historic building. Note that the demolition of a historic structure requires consent from the LPA and this is most unlikely to be given, nor is it normally desirable; most old buildings can be saved by restoration. As stated above, the total wall load has to be reduced, and perhaps one method of achieving this is to construct a strengthened wall plate around the whole building. Again this can be done discreetly so that any steel used for this purpose is not on general show.

Rising and Penetrating Damp

On most farm buildings there is a lack of a damp-proof course in the wall – this was never a requirement when the building was used for livestock, for example. One may be inserted, that is a chemical, mechanical, passive/active ion or internal vertical one. The most popular is chemical, but it should be borne in mind that this is ineffective in many situations, for example, for rubble fill, cob, stud or infill walls. In these situations the wall is a thick, solid construction and not a cavity

ABOVE: *A farm building converted into a restaurant, country store and vineyard. The extensive use of timber makes the building more appealing.*

BELOW: *The converted restaurant, store and vineyard; the use of plants and barrels helps to make the building fit in.*

Another building on the same site as in the previous plate, showing tasteful conversion.

recommended, unless they perform another purpose such as being an essential structural requirement; they also take up much space.

As an aside, the modern practice of painting external walls with a 'sealant', with the promise that this will exclude damp, is a total misnomer. The wall must be allowed to breathe; moisture travels through the wall, or at least it should do, all the time. If the outer surface is coated with a layer that will not allow moisture through, the result will be a breakdown in the very fabric that you were trying to protect.

Insulation

By referring to the current Building Regulations, the requirement is for a U-value of 0.45W/sq mK (or 0.25 for pitched roof with loft void) for all normal elements of an industrial building. This level of insulation is usually obtainable with a dry lining system, but it is a problem if the walls are to remain exposed. Perhaps it may be treated as an area with low-level heating requirements; alternatively, a higher roof U-value could be provided to balance the lower wall value. It is often found that some local authorities enforce the rules strictly, while others will not permit any relaxation of the rules whatsoever. It is also true that the U-value given to the old, existing wall is at best a reasonable guess. There will always be a need to expose historic fabric of interest, and this should be sufficient reason to argue a case with the local authority.

In general terms, it is important to ensure that the building gives ideal working conditions, meaning a high insulation value combined with fully controllable, natural ventilation is necessary. The

wall, as modern structures would be. It is possible in these instances to allow the wall to breathe with a dry lining on the interior and ensuring that the wall can breathe on the outside. With this system penetrating damp is dealt with together with the insulation requirements.

The common practice of putting external render on to a wall not only hides the historic cladding but it is also likely to accelerate the deterioration of the building fabric, particularly timber framing. The application of surface, silicone-type treatments is not cheap, especially for the short-term benefits, but the maintenance of jointing is imperative. In some circumstances, for example, in severe exposure or very weak surfaces, external cladding made from lapped boarding gives the air movement and protection required.

The most favoured solution is usually to provide an internal dry lining of insulated plaster board, mounted on plaster dabs or firing pieces if the wall is not very flat. Internal cavity walls are not really

actual insulation system that is used together with the ventilation are very important, bearing in mind the increased risk of interstitial condensation (that is, condensation occurring within the insulation material itself), and the need for the building fabric to breathe.

Natural Light and Ventilation

These topics should be considered in conjunction with the external fabric and are often a cause of conflict between the building and the planning control department (often situated in different locations). Interestingly, there is no minimal requirement for natural light under the Building Regulations; the often quoted rule of thumb for housing was 20 per cent of the floor area for windows of which 10 per cent should be opening for ventilation (Coates, 2001). The problem with this was that it led to so many openings that the original structure was virtually destroyed. Some basic guidelines therefore may be given:

- glazing should reduce glare and look dark; a good example would be to use Austin Bronze or similar
- windows could also be required to duplicate as fire escapes
- frames ought to be painted black to reproduce the tarring effect
- windows should be placed so that there are ventilation options to suit the conditions; they must be easy to access and operate and should also provide for background ventilation without being a compromise to security
- roof lights should be avoided where possible; if they are used they should go on the northern face and utilize dark, anti-sun glass; the use of roof lights can easily spoil a farm conversion and

there are many examples of where this has been done
- for enhanced air movement in the summer, it is best to insert passive background vents through the roof to enhance the stack effect and/or high level mechanical extraction
- air conditioning ought to be avoided through the design, if possible.

Highway Access

It is usually recommended that the working farm and the diversification enterprise are kept entirely separate; this might require some relocation. A shared main entrance may be acceptable, provided that it is kept reasonably clean; an entrance/exit that is continually dirtied by cows, for example, would not be acceptable. Interestingly, B1 (offices) use does not limit the size of vehicle access, and many farmers are having to accommodate 40-tonne lorries on their farms. Development and the design within the site can accommodate whatever lorry size is applicable, but the highway itself might be totally inadequate; planning approval would not be expected in such cases, and often any decisions are marginal. It is possible to avoid such conflicts by the farmer's imposing letting restrictions. Farms that are in remote locations should not expect approval, unless the application is put in in partnership with a suitable local business.

Parking and Turning Space

It is important to provide adequate loading and turning space in a farm conversion scheme – bare land is cheap and plentiful in most farm schemes. The general recommendation would be to provide a well-designed loading and turning area, probably surfaced in tarmac, and a large, landscaped car park with the right fall on it laid in scalpings over Terram fabric (Coates, 2001). It is well to remember that the number of employees, and therefore cars, will depend on the business, and so the main thrust has to be one of flexibility. The turning circle of a 40-tonne lorry is 18–20m (20–22yd) so this must be borne in mind and a similar space should also be provided for emergency vehicles. It is wise to restrict the access to any courtyard to delivery vehicles only; an area of 80m × 80m (88yd × 88yd) of permanent pasture with good access would give a helicopter room for landing, should this be required.

Water

The flow rates to farms are often lacking, not usually through the poor head but more often because of the small diameter pipes. Really the farm should have a 75mm (3in) internal diameter pipe, which will be sufficient for the fire hydrant also; if this is not feasible, then the pond in the farmyard might provide back-up. The main problem is caused by the water authority which insists on providing an individual service and meter to each 'property'. This means that it would usually renew the service from the nearest mains supply; it also implies some speculation when it comes to guessing the number of subdivisions required. The main distribution pipes should be laid in no less than 63mm (2.5in) MDPE pipe. It should be said that businesses that require water for industrial purposes are not usually suitable for rural locations because of the resulting effluent.

Electricity

Most conversions now put in a three-phase electricity supply because it is so widely used. However, this type of supply is expensive to install and the delay to put it in can be considerable, so it would be wise to talk to the electricity supplier very early in the project. It might even involve planning permission for any alterations to overhead services, so there is an even stronger reason to discuss it with the electricity company. Most distribution systems come in to a central source with meters and sub-mains to each unit on the farm.

Foul Water and Surface Water Drainage

Provided that there is no trade waste (such as copper- or aluminium-impregnated waste water), the quantity of effluent should be minimal, and so it is feasible to upgrade the farmhouse septic tank or to provide a small treatment tank. The Environment Agency may take up to five months to issue a 'discharge consent'; it is essential to keep this below 5cu.m/day to avoid the annual charges and the high standards required for volumes over 5cu.m (Coates, 2001).

As far as flash flooding is concerned, it is essential that the site has been investigated together with any old, clay land drain systems found and documented. If this discharges into a soakaway, it ought to be diverted if at all possible into open ditches. Any potential hazards such as courtyard tanks should be filled or secured in some way; clay drains are particularly susceptible to penetration and interference by tree roots, for instance.

CHAPTER 15

Safety Legislation

INTRODUCTION

Health and safety is a subject that often receives little or no recognition within the confines of a company. The necessary paperwork may be duly completed, procedures are put in place, and then everyone goes about their duty in a manner to suit themselves. Every year about fifty people are killed on farms and horticultural holdings and many more are injured by work activities. As well as the personal and social costs of accidents, which can be lifelong, the financial costs can include:

- sickness payments and recruitment/ training costs for replacement staff
- loss of output – key staff off work or temporary replacements not as effective
- inability to carry out weather-critical operations at the right time
- damage to machinery, buildings and products
- administration costs – investigating the incident, clearing up and repairs
- insurance and legal costs and adverse publicity.

All these costs can be avoided – thereby increasing profits as a result.

Although the text primarily concentrates on legislation that is relevant to British conditions, the general underlying principles are the same on farms anywhere in the world. The main piece of legislation in this country is the Health

and Safety at Work, Etc Act 1974, and the consequential regulations and codes of practice written beneath this main act in order to add extra help and guidance to users and operators.

WORKPLACE SAFETY AND WELFARE

Some important regulations are:

- The Workplace (Health, Safety and Welfare) Regulations 1992 aim to protect the health and safety of everyone in the workplace, and to ensure that adequate welfare facilities are provided for people at work
- The Confined Spaces Regulations 1997 set out precautions that must be taken before work is conducted in a confined space
- The Health and Safety (Safety Signs and Signals) Regulations 1996 require a safety sign where there is significant risk to health and safety not controlled by other methods
- The Lifting Operations and Lifting Equipment Regulations 1998 (LOLER)
- Provision and Use of Work Equipment Regulations 1998 (PUWER)
- The Work at Height Regulations 2005.

Many injuries in agriculture result from slips, trips and falls in workplaces such as buildings or yards. The risks of falling from a height are clear, so make sure that no one can fall more than 2m (6ft) from open edges, such as catwalks above grain bins or feed lofts. If the risk of injury is great from falls less than 2m, such as into a tank or pit or on to projecting objects, you must also take action. Guard rails at a maximum height of 1100mm (3.6ft) above the working surface or fencing will be suitable. Also check that working areas are free from obstructions, such as trailing cables, sacks or pallets, and that there is enough space for storing tools and materials. Keep your buildings in good repair, making sure that floors are not overloaded, especially in feed lofts or older buildings.

Provide:

- handrails on stairs and ramps where necessary and safety hoops or rest stages on long, vertical, fixed ladders (for instance, into grain bins) used regularly
- good drainage in wet processes such as vegetable-washing areas or dairies, and outdoor routes salted, sanded and swept during icy conditions
- adequate and suitable lighting; use natural light where possible but try to avoid glare, note that some fluorescent tubes flicker and can be dangerous, making rotating machinery appear stationary; well-lit outside areas will help security
- adequate temperature and ventilation, including fresh air in outdoor workplaces
- safety signs where a significant risk to health and safety remains after you have taken other control measures identified by your risk assessment.

The Work at Height Regulations 2005 state that, for any access to heights from which a fall is liable to cause injury, in circumstances where the access is not temporary, measures must be taken to provide safe access and to prevent falls from height, see later for more detail. (British Potato Council, 2005)

Workplaces can cause health problems, so make sure you provide:

- seats with a backrest supporting the small of the back and, if needed, a footrest, where work can be done seated, as for vegetable grading
- machine controls designed and arranged to provide a comfortable working position
- engineering controls, for instance, local exhaust ventilation systems to reduce health risks from noise or dangerous substances such as grain dust
- well-designed tools and working areas to reduce hand and forearm injury caused by repetitive movements, as on vegetable or fruit-grading lines.

Toilet and Welfare Facilities

There is a risk of illness from hazardous substances and from muck or other animal products carrying potentially hazardous micro-organisms. If you have full- or part-time, casual or permanent staff, provide rest facilities and:

- clean, well-ventilated toilets
- wash basins with hot and cold (or warm) water, soap and towels (or a hand dryer)
- portable washing facilities, or hand wipes, for workers working away from base
- changing facilities where special clothing is worn
- a clean drinking water supply (marked

to distinguish it from any non-drinkable supply).

Confined Spaces

A confined space is anywhere which, because it is enclosed, gives rise to a risk of serious injury from fire or explosion, loss of consciousness from lack of oxygen, drowning, or asphyxiation due to being trapped by a free-flowing solid. Confined spaces on farms are found in produce stores such as grain/forage silos and bins or controlled fruit and vegetable stores and pits such as grain elevator pits, slurry pits and chambers or vehicle-inspection pits. There have been deaths in confined spaces on farms; sometimes more than one person has been killed – the second person being a would-be rescuer. If you have areas which present any of these risks you must avoid working in the confined space if you can (can the work be done from outside?) and follow a safe system of work if you really have to work in a confined space. Consider the need for competent people, testing the atmosphere to make sure it can support life and does not contain dangerous levels of gases such as hydrogen sulphide – but remember that some areas such as slurry pits may continue to give off poisonous gases after testing; whether the area is adequately ventilated before entry, providing personal protective equipment (PPE), including breathing apparatus. And make arrangements in case something goes wrong – never enter the confined space without making proper emergency arrangements; rescue equipment, including harnesses and safety lines, should be provided, and ensure that you can rapidly notify the emergency services if something goes wrong.

Fire Precautions in Workplaces

Assess the risks if a fire were to break out, and make sure that:

- you have safe means of escape, kept free from obstructions and clearly marked
- everyone knows what to do if a fire starts, especially how to raise the alarm – display fire action instructions and have a fire drill periodically
- fire alarms work (check them weekly) and that they can be heard everywhere over normal background noise
- you have enough extinguishers, of the right type and properly maintained, to deal promptly with small outbreaks of fire make sure staff know how to use them.

PREVENTING FALLS

Falls from ladders, scaffolds and bale stacks are a major cause of serious accidents, many of which are fatal.

Ladders

Never work from a ladder if there is a safer way of doing the job, by using a scaffold or working platform, for example. If the job is quick and simple, you can use a ladder but always make sure it:

- has a level and firm footing (never use unsteady or slippery bases, for instance, oil drums, boxes, planks or an unclean animal yard)
- is not placed against a fragile surface such as fibre cement gutters – use a ladder stay or similar
- set at the most stable angle – a slope

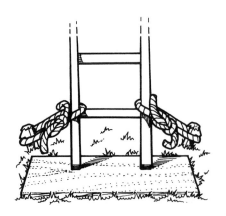

Ladder placed on a board to prevent it sinking into soft ground and tied to stop it from slipping. (HSE, 2002)

of four units up to each one out at the base
• extends at least 1m above the landing place or the highest rung in use, unless

there is a suitable handhold to provide equivalent support; extending ladders should overlap by at least three rungs
• is secured against slipping, for instance, by tying at the top, sides or foot; someone standing at the foot to prevent slipping is effective only with ladders less than about 6m (20ft) long; the figure above shows how a ladder might be secured at the foot.

But never use damaged or 'home-made' ladders – take them out of use and destroy or repair them, and do not place ladders where there is danger from moving vehicles, animals or electricity lines. Wear safety belts or harnesses during tree climbing or pruning or when gaining access to silos or bins without fixed ladders. The figure below shows how ladders should be positioned.

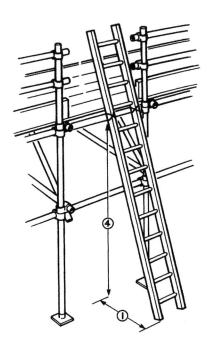

Ladders should be correctly angled, 'one out for every four up'. (HSE, 2002)

Scaffolds

Many tasks will be less hazardous if you do them from a properly designed and erected scaffold. Use competent and experienced workers to erect a scaffold, and make sure that they are under the control of, and are inspected regularly by, a competent person. Check that:

- the scaffold is placed on level, firm ground with baseplates and soleplates where necessary, properly braced with vertical supports (standards) every 2 to 2.5m (6–8ft)
- the platform is at least 600mm (24in) wide, with adequate supports, not more than 1.5m (5ft) apart
- scaffold boards are tied down or overhang each end support by 50 to 150mm (2–6in)
- you have provided fall protection such as guard rails, with the main guard rail at least 910mm (1yd) above the platform, with toe boards at least 150mm (6in) high. there should not be an unprotected gap of more than 470mm (19in) between the platform and any guard rail
- there is safe access to the scaffold – never climb the outside.

For mobile scaffolds, also:

- check the maximum recommended height in relation to the base dimension (including outriggers, if fitted). the base: height ratio is often 1:3
- tie them to the building or extend the base with outriggers, if using them outside in windy weather (always tie them if they are to be left unattended)
- clear the working platform of people and materials when the scaffold is being moved, move it only by pulling or pushing at the base

- wheel brakes must be 'on' and locked when the scaffold is used
- do not overload the working platform or apply pressure which could tilt the tower.

Working at Height

If access is required to, say, the top of boxes in a potato store, it is necessary to take measures to minimize any threat to an individual's safety when undertaking this task and to ensure compliance with the Work at Height Regulations 2005. These regulations require measures to prevent falls from heights liable to cause injury. The hazards can be separated into three main areas: access to the top of the stack; the risk of falls when on top of the boxes; and the removal of samples. Boxes should be stacked in large blocks wherever possible to reduce unprotected edges. Additional risk assessments will be required in stores where box layouts may be configured for separation of varieties and/or the use of tent drying systems and access is required to the top of the stack. These may involve reconsideration of traditional box layouts to reduce work at height risks.

Access
The Work at Height Regulations 2005 prohibits the use of portable ladders for regular access to crop at height, even if the ladder is fixed. Access to the top of box stores, if deemed necessary after risk assessment, can therefore no longer be made legally by the methods traditionally used in the industry. Access should therefore be made by one of the following methods:

- fixed ladders and staircases; it is important that, if used, these comply with relevant legislation, fixed access is difficult to manage if it is necessary to part load the store

- access towers and scissor-lifts: these systems can be expensive to buy and maintain as equipment has to be maintained in accordance with PUWER (Provision and Use of Work Equipment Regulations) and inspected in accordance with LOLER (Lifting Operations and Lifting Equipment Regulations); mobile access may be difficult to use when the store is full but is more flexible where it is necessary to part unload the store
- multiple-step short ladders: this simple, inexpensive system involves stepping boxes when loading to provide a safer means of access using short ladders which can be hung safely on a box; ladders should be attached to full boxes only and stay in place for the duration of storage; the system will provide access in existing box potato stores up to eight boxes high for a minimum loss of store capacity and a means for lowering samples to floor level; ladders should comply with guidance on ladder construction and maintenance; key design features to be considered include:
  - manufacture from robust material and methods
  - secure anchoring on centre or corner post(s) of the box
  - handholds above the box to provide ease of access to platform; if these form part of a barrier these must be a height of 1m above the box
  - sloping design of a pitch appropriate to box size
  - absence of protruding edges which may cause injury, for example, impaling in the event of a fall
- fixed platform: a more expensive solution which will meet legislative requirements and may be required in

all new stores; this means of access could be applied in existing stores with plenum chambers or suction wall systems but otherwise may be difficult to fit retrospectively; note that any retrospective fitment, either inside or outside the building, may require inspection and approval (British Potato Council, 2005).

BUILDING WORK

Most activities involving structural work on farms are subject to the Construction (Health, Safety and Welfare) Regulations 1996 and the Construction (Design and Management) Regulations 1994 (CDM), and where they apply it is required that health and safety is managed throughout all stages of a project, from conception, design and planning through to site work and subsequent maintenance and repair of the structure. Most farms carry out some building work, from dismantling and re-erecting entire buildings to repairing fragile roofs. All such work involves risks – the construction industry is generally accepted to be just as hazardous as agriculture – and so you must put proper controls in place. CDM will apply to construction where the work will last for more than thirty days and involve five or more people on site at any one time. In these cases you, as the 'client', will have legal duties to comply with and thus you should obtain advice.

Working on Fragile Roofs

Most types of fibre cement roofs (commonly known as 'asbestos' roofs, but not always containing asbestos) will be fragile. Roof lights will often also be fragile. Remember that it is never safe to walk across any fragile roof without using roof ladders or crawling boards. Always consider first whether it is really necessary to have access to the roof – does the work need to be done, or could it be done in some other way such as from an elevated work platform? If you, your employees or contractors repair, replace or clean roofs or need access to them for inspection or to get to plant, follow these rules:

- make sure everyone knows the precautions to be followed when working at heights
- fix a prominent permanent warning notice at the approach to any fragile roof
- never walk on fragile materials such as asbestos or other fibre cement sheet, many roof lights or glass (beware – roof lights and glass may have been painted over)
- never 'walk the purlins'
- roof ladders or crawling boards must span across at least three purlins; they should be at least 600mm (2ft) wide and more when the work requires it
- do not use a pair to 'leapfrog' across a fragile roof – provide enough boards
- take precautions to prevent anyone from falling from the ladder or board – edge protection or safety harnesses or safety netting is necessary where this is not feasible; take special advice – but remember that harnesses require adequate attachment points and rely on user discipline and training to ensure that they are consistently and correctly used
- roof ladders must be securely placed, with the anchorage bearing on the opposite roof; do not rely on the ridge caps or tiles for support since they can easily break away; never use gutters to support any ladder.

Working on or Passing Near to Fragile Roofing Material

You will need to provide protection when anyone passes by or works nearer than 2m (6ft) to fragile materials, for example, during access along valley gutters in a fragile roof, when an otherwise non-fragile roof contains fragile roof lights or during access to working areas on a fragile roof. The figure overleaf shows detail of a roof protection system. You should wherever possible make sure that all fragile materials 2m or closer to the people at risk are securely covered, or provide full edge protection (top rail, intermediate guard rail or equivalent and toe board) around or along the fragile material to prevent access to it, and make sure you take precautions when installing such protection, for instance, use netting. If it is not reasonably practicable to provide such protection use safety nets or harnesses, but make sure staff are trained and competent in their installation and use.

Excavation

If excavating remember:

- trench sides may collapse suddenly whatever the nature of the soil
- you need to decide before carrying out the work what precautions will be required to protect against collapse of the sides, for instance, shoring or battering
- keep a clear area around excavations to prevent people, materials or vehicles falling in, and the weight of soil or equipment from causing the sides to collapse
- if you need to enter the excavation, provide safe access

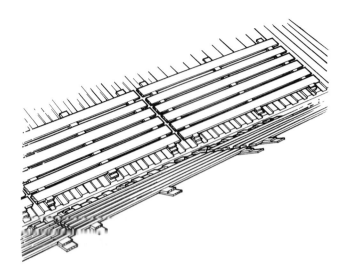

Permanent protection installed at the valley gutter on a fragile roof. (HSE, 2002)

- there may be poisonous or asphyxiating gases in sewer openings, from marshy ground or confined spaces.

Keep well away from overhead electricity lines and underground services, including cables and gas pipes – consult the utility companies before you start work to establish where the lines and pipes are (HSE, 1/02).

ASBESTOS

The Control of Asbestos at Work Regulations 2002 require employers to prevent the exposure of employees to asbestos. If this is not reasonably practicable the law says that their exposure should be controlled to the lowest possible level. Before any work with asbestos is carried out, the regulations require employers to make an assessment of the likely effects of any exposure. The assessment should include a description of the precautions to be taken to control dust release and to protect workers and others who may be affected by that work. If you are employing a contractor to work on your premises, make sure that either the work will not lead to asbestos exposures or that the contractor has carried out this assessment and identified work practices to reduce exposures (HSE, 12/04). Asbestos was used widely in agriculture as lagging on plant and pipework, in insulation products such as 'fireproofing' panels and in asbestos-cement roofing sheets. Although in new products it has been replaced by other materials, many older ones on farms may contain asbestos. Exposure to asbestos fibres as a result of disturbing the fabric of a building can be fatal (HSE, 1/02).

Breathing in contaminated air can lead to asbestos-related diseases, mainly cancers of the lungs and chest lining. Asbestos is a risk to health only if its fibres are released into the air and breathed in. Past exposure to asbestos currently kills 3000 people a year in Great Britain. This number is expected to go on rising for the next ten years. There is no cure for asbestos-related diseases and there is usually a long delay between first exposure to and the onset of disease; this can vary from 15 to 60 years. Only by preventing or minimizing these exposures now will asbestos-related disease eventually be wiped out.

There are three main types of asbestos still found in premises, these are commonly called: 'blue asbestos' (crocidolite), 'brown asbestos' (amosite) and 'white asbestos' (chrysotile).

All are dangerous, but blue and brown asbestos are more hazardous than white. You cannot identify them just by their colour. Although it is now illegal to use asbestos in the construction or refurbishment of any premises, many thousands of tonnes of it were used in the past and much of it is still in place. As long as it is in good condition and not being or going to be disturbed or damaged there is no risk. But if it is disturbed or damaged it can become a danger to health because its fibres are released into the air and they can be inhaled.

Where Is Asbestos Found in Buildings?

You are most likely to come across asbestos in the following materials:

- sprayed asbestos and asbestos loose packing – generally used as fire breaks in ceiling voids
- moulded or preformed lagging – generally used in thermal insulation of pipes and boilers
- sprayed asbestos – generally used as fire protection in ducts, firebreaks, panels, partitions, soffit boards, ceiling panels and around structural steel work
- insulating boards used for fire protection, thermal insulation, partitioning and ducts

- some ceiling tiles
- millboard, paper and paper products used for insulation of electrical equipment; asbestos paper has also been used as a fire-proof facing on wood fibreboard
- asbestos cement products, which can be fully or semi-compressed into flat or corrugated sheets; corrugated sheets are largely used as roofing and wall cladding; other asbestos products include gutters, rainwater pipes and water tanks
- certain textures coatings
- bitumen roofing material
- vinyl or thermoplastic floor tiles.

How Can You Dispose of Asbestos?

Asbestos waste, whether in small amounts or large-scale quantities removed by contractors, is subject to waste management controls set out in the Special Waste Regulations 1996. Asbestos waste should be double-bagged in heavy duty polythene bags and clearly labelled as prescribed for asbestos before it is transported to a disposal site. The waste can be disposed of only at a site licensed to receive it. Your local authority will have information on such sites in the area.

Checklist

Find   you must check whether materials containing asbestos are present

Condition you must check what condition the material is in

Presume   you must assume the material contains asbestos unless you have strong evidence that it does not

Identify   if you are planning to have maintenance or refurbishment of the building carried out or the material is in poor condition, you may wish to arrange for the material to be sampled and identified by a specialist

Record   record the location and condition of the material on a plan or drawing

Assess   you must decide whether the condition or the location means the material is likely to be disturbed

Plan   prepare and implement a plan to manage these risks

Deciding what to do

Good condition
- the general condition of the material should be monitored at regular intervals
- where practical, the material should be labelled
- inform contractor and any other worker likely to work on or disturb the material

Minor damage
- material should be repaired and/or encapsulated
- condition of the material should be monitored at regular intervals; where practical, the material should be labelled
- inform contractor and any other worker likely to work on or disturb the material

Poor condition
- asbestos in poor condition should be removed

Asbestos disturbed
- asbestos likely to be disturbed should be removed (HSE, 12/04)

ASSESS THE RISKS

You are legally required to assess the risks in your workplace. A risk assessment must be carried out for all roof work. It is essential that risks are identified before work starts and that necessary equipment, appropriate precautions and systems of work are provided and implemented. The process of preparing an assessment for the jobs covered by this section should be straightforward and relatively quick.

Method Statements

Except for the simplest jobs where the necessary precautions are straightforward and can easily be repeated (such as to use a proper roofing ladder to replace a ridge tile), you should prepare safety method statements relating specifically to the job in hand. They should describe clearly the precautions and systems of work identified during the assessment. Everyone involved in the work needs to know what the method statement says and what he has to do – if he cannot understand the precautions or systems needed then he should not be permitted to carry out the work. Make sure you have arrangements for supervision during the work to check that the laid down procedures are followed.

The method statement should cover the following areas:

- getting on and off the roof: safe access is essential, a properly secured ladder is the minimum requirement

- edge protection: wherever anyone could fall more than 2m (6ft) the first line of defence is to provide adequate edge protection. the minimum requirements are:
  - a main guard at least 910mm (36in) above the edge
  - a toe board at least 150mm (6in) high
  - an intermediate guard rail or other barrier so that there is no gap more than 470mm (18.5in)
- protection against falling through fragile materials – adequate supports or covers must be used
- reducing the need for workers to move about the roof, for instance, by arranging for the right materials to be lifted to the right place at the right time.

Make sure that appropriate warning signs are displayed on existing roofs, particularly at roof access points.

Fall Arrest Equipment (Nets and Harnesses)

In some cases you may have to work from a crawling board or staging (perhaps when removing a sheet) or next to an opening (perhaps created by removing the roof sheet) without adequate edge protection. If there is a risk of falling more than 2m from the crawling board or staging through the roof or through the opening, you will need to provide a safety net immediately beneath the roof. This will prevent anyone from falling through the roofing material or the opening you created (for example, to replace each sheet). Safety nets should not be installed by people who are not competent to do so, such as most farm staff. You may therefore here need to take independent advice or help. Safety harnesses are not suitable for use by people who have not been trained in their use nor without constant supervision by a competent person. Unless you are confident that you have the right staff, equipment and systems of work you should not use harnesses.

Training

Some people are not suited to work at heights and could put others at risk, for instance, if they suffer vertigo they should not be asked to do this type of work. Those who are suitable need appropriate knowledge, skills and experience to work safely, or must be under the supervision of someone else who has. They need to be able to recognize the risks, understand and follow safe systems of work and be competent in skills such as installing edge protection and operating a mobile access platform. Training will usually be required to achieve competence. It is insufficient to hope that workers will 'pick up' safety on the job. Consult a training provider such as an agricultural college or a training group (HSE, 2004).

APPENDIX

Planning Application Forms

There follows some examples of application forms which may be used by members of the public when they are proposing new developments.

pp.179, 180: Example planning application form used by Chelmsford Borough Council. (Chelmsford Borough Council, 2001)

p.181: Example full plans submission form used for building regulation approval used by Chelmsford Borough Council. (Chelmsford Borough Council, 2001)

p.182: Example building notice form, as used by Chelmsford Borough Council. (Chelmsford Borough Council, 2001)

p.183: Example application for listed building/conservation area consent form as used by Chelmsford Borough Council. (Chelmsford Borough Council, 2001)

PLANNING APPLICATION

**APPLICATION FOR PERMISSION TO CARRY OUT DEVELOPMENT
UNDER THE TOWN & COUNTRY PLANNING ACT (1990)**

Town Planning Services

**SIX COMPLETED COPIES OF THIS FORM AND SIX SETS
OF PLANS MUST BE SUBMITTED
PLEASE READ "NOTES FOR GUIDANCE" BEFORE
COMPLETING YOUR APPLICATION**

For Office use only	**FORM T.P.1**
Ref:	

---

1. Applicant (in block capitals)

    Name ....

    Address ....

    ....

    ....

    Post Code ....Tel. No. ....

    Agent (if any) to whom correspondence should be sent (in block capitals)

    Name ....

    Address ....

    ....

    ....

    Post Code ....Tel. No. ....

---

2. Full address or location of the land to which
    this application relates (**edged in red** on
    the accompanying site plan)
    ....
    ....
    ....

---

3. State brief particulars of proposed development
    including the purpose(s) for which the land
    and /or buildings are to be used, and including
    changes of use.
    ....
    ....
    ....

    Note: If the application is for industrial, office, warehousing, storage or shopping purposes, you must also complete Form T.P.2

---

4. Site Area (metric)

---

5. State whether applicant owns or controls any adjoining land and if so, show location
    (**edged in blue**) on the accompanying site plan   YES/NO

---

6. State whether the proposal involves: YES/NO

    (a) New building(s)...

    (b) Alteration or extension..

    (c) Change of use of land or building(s).........................
    (d) Construction of new   Vehicular........
         access to a highway   Pedestrian.......
    (e) Alteration of an existing   Vehicular........
         access to a highway   Pedestrian.......
    (f) Stopping up or Diversion of a
         public right-of-way   ....

    (g) Other...

    If residential development, state number of dwelling units
    proposed and type:

    } Point of access must be
      indicated on submitted plans

    If YES, indicate on site plan

---

7. State number of residential units lost through demolition or
    change of use (if applicable)

---

8. Does the proposed development affect a building included in a
    List of Buildings of Special Architectural or Historic Interest or
    involve the demolition of a building in a conservation area.   YES/NO

---

9. Particulars of Present and Previous Use of Building or Land.

(a) Present use of building/land ....

(b) If vacant, the last previous use and date that use ceased. ....

---

10. State whether application is for:-

(a) Outline Planning Permission?

If you are applying for outline planning permission please tick which of the following matters (if any) are part of this application:-

☒Siting   ☐Landscaping   ☒Means of Access

☐Design   ☒External Appearance

(b) Full planning permission including erection of buildings, extensions and changes of use?

(c) Approval of reserved matters following the grant of outline permission?

If answer to (c), (d) or (e) is YES, state date and reference number of previous permission:

(d) Renewal of temporary or time limited permission?

....

(e) Continuance of use without complying with a condition subject to which planning permission has been granted?

If a Building Regulation application has been Submitted prior to the planning application please give ref. number and date below

....

---

11. Is the application seeking to retain or regularise an existing use or development?   YES/NO

---

12. Landscaping
(a) Are there any trees or hedges on the site or along its boundaries?   YES/NO   **If YES, indicate position on plans.**

(b) Do you intend to lop, top or fell any trees or hedges?   YES/NO
     If YES, give precise particulars indicating positions on submitted plans.

---

13. Drainage
What method is proposed for:-   (a) disposal of surface water? ....

                                       (b) dealing with foul sewage? ....

---

14. Materials
Give details (unless the application is for outline permission) of the colour and type of external materials to be used, if known.

(a) Walls ....

(b) Roof ....

(c) Boundary walls and/or fences ....

---

15. List of drawings and plans submitted with the application

---

16. Preliminary Discussions: If you have had previous discussions or correspondence with the Council regarding this proposal please give the name of the offi.cer and any reference number quoted by the Council.

---

17. State type of Ownership Certificate submitted with this application (A or B)

---

18. ADDITIONAL INFORMATION which you may wish to give in support of the application should be submitted separately (four copies please).

I/We hereby apply for permission to carry out the development described in this application and the accompanying plans, and in accordance therewith.

Date .... Signed ....

on behalf of ....
(Insert applicant's name if signed by an Agent)

BOROUGH COUNCIL

DEVELOPMENT SERVICES GROUP
Building Control Services
Tel: 01245 606431
Civic Centre Duke Street Chelmsford Essex CM1 1JE

Two copies of this form are to be filled in by the person who intends to carry out building work or agent and be accompanied by two sets of plans. Please type or use block capitals. If the form is unfamiliar please read the "Notes for Guidance" or consult the office indicated above.

FULL PLANS SUBMISSION

The Building Act 1984
The Building Regulations 2000

Building Regulations
Ref Number:

---

1 Applicant's details
Name: _____
Address: _____
_____
_____ Postcode: _____ Tel: _____

2 Agent's details (if applicable)
Name: _____
Address: _____
_____
_____ Postcode: _____ Tel: _____

3 Location of building to which work relates
Address: _____
_____
_____ Postcode: _____ Tel: _____

4 Proposed work
Description: _____
_____
_____
_____

5 Use of building
1 If new building or extension please state proposed use: _____
2 If existing building state present use: _____
3 Number of storeys in existing building: _____

6 Conditions (see note 4)
Do you consent to the plans being passed subject to conditions where appropriate?   YES/NO
Do you consent to an extension of the statutory time limit by 3 weeks if necessary?   YES/NO

7 Charges (see Guidance Note on Charges for information)
Plan charge £ _____ plus   VAT £ _____ = Total £ _____
Estimated cost of works (where applicable) £ ....

8 Additional Information
1 Foul water
2 Surface water ....
3 Nature of subsoil
4 Are there are any trees on, or adjacent to, the site?   YES/NO If yes, show details on plan

9 Town Planning
Has a town planning application been submitted for this proposal?   YES/NO If yes, reference number: _____

10 Statement
This notice is given in relation to the building work as described, is submitted in accordance with regulation 12(2)(b) and is accompanied by the appropriate charge. I understand that a further charge will be payable following the first inspection by the local authority.
Name: _____ Signature: _____ Date: _____

BUILDING CONTROL

IMPORTANT NOTICE
A Building Notice <u>cannot</u> be used for works that are within 3 metres of a public sewer, as shown on the AWS Public Sewer records.
Please telephone 01245 606431 if you require further information.

BUILDING NOTICE

The Building Act 1984
The Building Regulations 1991

Chelmsford
BOROUGH COUNCIL

DEVELOPMENT SERVICES GROUP
Building Control Services

Civic Centre Duke Street Chelmsford Essex CM1 1JE

Building Regulations Number:

This form is to be filled in by the owner or agent. Please type or use block capitals.
If the form is unfamiliar please read the "Notes for Guidance" or consult the office indicated above.

1 Applicant's details
Name: _____
Address: _____
_____
_____ Postcode: _____ Tel: _____

2 Agent's details (if applicable)
Name: _____
Address: _____
_____
_____ Postcode: _____ Tel: _____

3 Location of building to which work relates
Address: _____
_____
_____ Postcode: _____ Tel: _____

4 Proposed work
Description: _____
_____
_____
_____

5 Use of building
1 If new building or extension please state proposed use: _____
2 If existing building state present use: _____
3 Number of Storeys in existing building: _____

6 Commencement
Date if known (48 hour notice required) _____

7 Charges (see Guidance Note on Charges for information)
Building Notice charge £ _____ plus VAT £ _____ Total £ _____
Estimated cost of works (where applicable) £ ....

8 Additional Information
1 Foul water
2 Surface & Roof water

9 Town Planning
Has a town planning application been submitted for this proposal? YES/NO If yes, reference number:

10 Statement
This notice is given in relation to the building work as described, is submitted in accordance with regulation 11(1)(a) and is accompanied by the appropriate fee.
Name: _____ Signature: _____ Date: _____

LA *Local Authority*
BUILDING CONTROL

APPLICATION FOR LISTED BUILDING/
CONSERVATION AREA CONSENT

PLANNING (LISTED BUILDINGS AND CONSERVATION AREAS) ACT 1990
TOWN AND COUNTRY PLANNING (LISTED BUILDINGS AND BUILDINGS
IN CONSERVATION AREAS) REGULATIONS 1990

Town Planning Services

PLEASE SUPPLY FOUR COMPLETED COPIES OF THIS FORM AND FOUR SETS OF PLANS PLEASE READ "NOTES FOR GUIDANCE" BEFORE COMPLETING YOUR APPLICATION	For Office use only /CHL/ Ref

1.

Applicant (in block capitals)	Agent (if any) to whom correspondence should be sent (block capitals)
Name	Name ....
Address	Address ....
....	....
....	....
Post Code .... Tel. No.	Post Code .... Tel. No. ....

2.

Full address or location of the building(s) to which this application relates (the site should be edged in red on the accompanying site plan with any other land within the ownership or control of the applicant edged in blue)

....

....

3. Type of Building (eg) dwelling house, barn .... **5.** Is the building included in the Statutory List as a Grade I, Grade II or Grade II* building (see note below)	**4.** Has a financial grant been made or applied for in respect of the building (other than an Improvement Grant under the Housing Acts)? If so, by or to whom and when? ....

6.

Full description of the proposed works/development, including types of materials to be used. If the proposal is to demolish, please indicate the extent of the demolition ....

....

....

....

....

....

I/We hereby apply for permission to carry out the works/development described in this application and the accompanying plans, and in accordance therewith.

Date .... Signed ....

on behalf of ....
(Insert Applicant's name if signed by an Agent)

Only buildings of Grade I, II and II* are included in the Statutory List prepared by the Secretary of State under section 1 of Planning Listed Building and Conservation Area Act 1990 and these are therefore the only ones covered by the Listed Building Regulations.

Glossary

AONB area of Outstanding Natural Beauty

Anaerobic without oxygen

Bargeboard sloping roof trim of normally wood fixed in pairs along the gable to cover the roof timbers

BSP British standard pipework

Cob clayey soil and straw, shaped in place to make the walling of an earth building

CCTV closed circuit television

DPC damp-proof course

Eaves lowest part of a sloping roof; eaves may have a horizontal fascia which carries the gutter

EIA environmental impact assessment

EU European Union

Flue pipe to the outside from a boiler or heater, to take away poisonous gases

Gable triangular part of the end wall of a building

GGBS ground granulated blast-furnace slag

LCT lower critical temperature

LPA local planning authority

Lintel beam over a door or window, usually carrying a wall load

Mitre one of two matching ends, each cut off at half the bend angle (usually 45 degrees) to make a mitred joint

Monopitch a pitched roof that slopes in only one direction, as in a lean-to

MDPE medium density polyethylene

NFU National Farmers Union

Oversite concrete layer of (dry lean) concrete about 100mm [4in] thick required to seal the soil under the ground floor of a house

Over-sail to overhang on a building (usually the roof)

Oast house building containing ovens for drying hops

OPC ordinary Portland cement (normally now called PC)

PFA pulverized fuel ash

Potable water drinking water

Portal frame frame of two columns with either a horizontal beam between them or two sloping rafters, may be of steel, concrete or timber

Purlin one of the horizontal roof beams parallel to the eaves and ridge, carried on the main framing members such as trusses

PTFE polytetrafluoroethylene

PVC polyvinyl chloride

QA quality assured

RMC Ready Mix Concrete

RHPC rapid hardening Portland cement

Render usually the mortar used as an undercoat on a wall

Spaced (Yorkshire) boarding vertical attachment of boards to the side/ends of a building that have spaces between them for ventilation

SRPC sulphate-resisting Portland cement

Slag glassy by-product from steelmill blast-furnaces; it is granulated by cooling with water and then ground to the fineness of cement

TPO tree preservation order

UCT upper critical temperature

Verge sloping edge of a pitched roof above a gable, can be flush or have an overhang

References

ADAS (1983), 'Insulation of Farm Buildings' (Leaflet 851)

ADAS (1991), 'Farm Diversification: new opportunities for profit'

ADAS (1992), 'Planning Dairy Units', FRBC

ADAS (1992), 'Planning Sheep Housing', FRBC

Barnes, M. and Mander, C. (1991) *Farm Building Construction* (Farming Press)

BRE (1993), 'Good Practice Guide 61. Design Manual, Energy Efficiency in Advance Factory Units'

Brent, Gerry (1991), *Housing the Pig* (Farming Press)

Brinkley, Mark (2002), *The Housebuilder's Bible*, 5th edn (Rodelia Ltd)

British Cement Association (1991), 'Advice on the safe use of Portland cement'

British Cement Association (1995), 'Site sampling and testing of concrete'

British Potato Council (2005), 'Health and Safety in Potato Stores – a guide to legislative requirements and best practice recommendations'

British Steel (Corus) Strip Products (1992), 'The use of profiled steel sheeting for farm buildings' (Farm Building Guide)

Bruce, J.M. (1987), 'The environmental requirements of livestock', IAgrE *Agricultural Engr*, (Winter 1987)

British Standard (1991), 'Buildings and Structures for Agriculture, Pt 40, Code of Practice for Design and Construction of Cattle Buildings'

Brunskill (1999), 'Traditional Farm Buildings of Britain and Their Conservation' (BAS Printers)

Carr, J. (1994), *Pig Farming* (May, June, July and Aug.)

Cemex Readymix (2005), 'Shaping the Future in Agricultural Concrete'

Cemex (2000), RMC Agricultural Legislation Guide

Chelmsford Borough Council (2001), Planning Application and other forms

Chelmsford Borough Council, Local Authority Building Control (1999), 'Why do you need Building Regulations Approval?'

Clark, J.A. (1981), 'Environmental Aspects of Housing for Animal Production' (Butterworths, London)

Coates, D. (2001), *RDBA Journal – Countryside Building*, vol. 1(2)

College of Estate Management (1993), *Rural Buildings*, vol.1

Cory, Bill (1991), 'Fans for today's agriculture', *Agricultural Engr*, Spring

Cull, S (1987), Design Guidance for Farm Buildings, Farm Buildings and Engineering (3), 3, 1987

DEVI (2006), sales literature

Department of the Environment (1993), Fuel Efficiency Booklet 7: Degree Days. Best Practice Programme

Department of the Environment, Transport and the Regions (1994), Planning Policy Guidance (PPG 7), The countryside – environmental quality and economic and social development

Electricity Association Services (1993), 'Electricity and You. How to Read Your Meter'

Electricity Council (1981), 'How to Read Your Meter'

Energy Efficiency (2000), 'The designer's guide to energy efficient buildings for industry', GPG 303

Farm Electric (1990), 'Controlled Environments for Livestock'

Farm Electric (1996), 'A Guide to Lighting Systems within Agriculture', Technical Note TN 63

Farm Buildings Centre, (unknown title, [n.d.])

Farmplus Construction Ltd (2005), sales literature

Farmers Weekly (30 Dec. 1994), 'Light programme'

Farmers Weekly (2 Feb. 2001), 'Don't get lost in the planning jungle'

FRBC [n.d.], *Farm and Rural Buildings Pocketbook*

Haines and Davies (1987), *Diversifying the Farm Business. A practical guide to the opportunities and constraints* (BSP Professional Books)

Hall, F. (1988), *Essential Building Services and Equipment* (Heinemann)

Hepworth Drainage (2000), sales literature

HSE (2002), 'Farmwise – your essential guide to health and safety in agriculture'

HSE (2003), 'General access scaffolds and ladders' (HSE Information Sheet)

HSE (2004), 'A short guide to managing asbestos in premises'

HSE (2004), 'Preventing falls from fragile roofs in agriculture' (HSE Information Sheet)

JS Industrial Services Ltd (2005), psychrometric chart

Livestock Systems (1997), *Joints in Concrete* (Simon Pearson)

Marley Eternit (1991), *The Farmer's Building Handbook*

Marley Eternit (2004), *Profiled Sheeting – choosing and using*

Maywick (2004), sales literature

Milk Development Council (1998), 'Calf Rearing – how to get it right and minimise losses'

McMullan, R. (2002), *Environmental Science in Building* (5th edn, Palgrave)

National Polytunnels Ltd (1997), sales literature

NFU (1994), 'Farming and Planning Law'

North Pennines Area of Outstanding Natural Beauty Steering Group (1998) (Eden District Council), 'Agricultural Buildings Design Guide'

Noton, N. (1982), 'Farm Buildings' (College of Estate Management)

Office of the Deputy Prime Minister (2003), 'Planning – A Guide for Householders. What you need to know about the planning system'

Payless DIY (1986), sales literature

Philips Lighting (1984), 'Artificial lighting in horticulture'

Powell-Smith and Bilkington (1995), *The Building Regulations – explained and illustrated* (10th edn, Blackwell Science)

Prag, P. (2000), 'Rural diversification', *Estates Gazette/Farmers Weekly*

Proctor (1997), sales literature

Pyramid (1996), sales literature

Riverina (1995), sales literature

Sainsbury, D. and P. (1988), *Livestock Health and Housing* (Bailliere Tindall)

Sainsbury, D. (1992), *Poultry Health and Management* (3rd edn, Blackwell Scientific Publications)

SAC [n.d.], 'Automatically Controlled Natural Ventilation'

Shufflebottom Ltd. (2005), example drawings

Titan Pollution Control (2004), sales literature

Watkins and Winter (1988), 'Superb Conversions? Farm diversification – the farm building experience' (CPRE)

Index